Higher Education

Increasing corporate social responsibility demands professionals possess the necessary knowledge, abilities, and competencies to answer the needs of a diverse organization's stakeholders. This book highlights the most recent issues related to higher education in the fields of management and engineering. It explains why a sustainable education is a requirement for professionals, as well as the organizations they collaborate with.

Higher Education: Progress for Management and Engineering focuses on the latest research findings in the field of higher and sustainable education. It discusses the progress, shares knowledge and insights on an international scale, and highlights the challenges faced to obtain and secure a more responsible and sustainable management system. Selecting different options and strategies, how to set priorities on managing competition, and how to succeed as an organization that can lead to successes in both national and international markets are covered within this book.

This book can be used as a reference for researchers, academics, managers, engineers, and other professionals involved in higher and sustainable education in management and engineering.

Higher Education and Sustainability

J. Paulo Davim, Professor, Department of Mechanical Engineering, University of Aveiro, Portugal

This new series fosters information exchange and discussion on higher education for sustainability and related aspects, namely academic staff and student initiatives, campus design for sustainability, curriculum development for sustainability, global green standards: ISO 14000, green computing, green engineering education, index of sustainability, recycling and energy efficiency, strategic sustainable development, sustainability policies, sustainability reports, etc. The series will also provide information on principles, strategies, models, techniques, methodologies, and applications of higher education for sustainability. It aims to communicate the latest developments and thinking as well as the latest research activity relating to higher education, namely engineering education.

Higher Education and Sustainability
Opportunities and Challenges for Achieving Sustainable Development Goals
Edited by Ulisses Manuel de Miranda Azeiteiro and J. Paulo Davim

Designing an Innovative Pedagogy for Sustainable Development in Higher Education
Edited by Vasiliki Brinia and J. Paulo Davim

Higher Education: Progress for Management and Engineering
Edited by Carolina Machado and J. Paulo Davim

For more information about this series, please visit: https://www.crcpress.com/Higher-Education-and-Sustainability/book-series/CRCHIGEDUSUS

Higher Education

Progress for Management and Engineering

Edited by

Carolina Machado
J. Paulo Davim

CRC Press
Taylor & Francis Group
Boca Raton London New York

CRC Press is an imprint of the
Taylor & Francis Group, an **Informa** business

First edition published 2023
by CRC Press
6000 Broken Sound Parkway NW, Suite 300, Boca Raton, FL 33487-2742

and by CRC Press
4 Park Square, Milton Park, Abingdon, Oxon, OX14 4RN

© 2023 Taylor & Francis Group, LLC

CRC Press is an imprint of Taylor & Francis Group, LLC

ISBN: 978-0-367-44418-1 (hbk)
ISBN: 978-1-032-29851-1 (pbk)
ISBN: 978-1-003-02123-0 (ebk)

DOI: 10.1201/9781003021230

Typeset in Times
by KnowledgeWorks Global Ltd.

Contents

Preface

Higher Education: Progress for Management and Engineering highlights the most recent issues related to higher education in the field of management and engineering. More and more corporate social responsibility demands from the different professionals, namely, managers and engineers, the necessary knowledge, abilities and competences in order to better answer to the needs of the diverse organization' stakeholders. A sustainable education is thus understood as a required issue that these professionals, as well as the organizations with which they collaborate (work), need to obtain.

Higher education is a critical tool to the organization development due to its ability to change and to drive change and progress in organizations. It contributes to organizations more knowledge-based which lead to a more effective sustainable development of all those who are part of the organization. Given its critical role, it is important to follow and understand the changes and progress that higher education has been facing. Particularly in the field of management and engineering, it is important to understand the progress of higher education as only in this way it will be possible to have sustainable and socially responsible organizations able to answer and, above all, to anticipate the demands of different stakeholders.

Higher education assumes a relevant role as it is vital to build the future of the society. It is the basis to obtain new skills, knowledge as well as new ideals necessary to an effective sociocultural and economic development. To understand and follow its progress is a concern for all those that are interested in being updated with the most recent knowledge. At the organizational level, it is critical that all its professionals, with a particular emphasis to managers and engineers, possess the necessary knowledge to guide the organization to success. This is why *Higher Education: Progress for Management and Engineering* is of huge importance for all those who somehow have interests in the organization, namely, researchers, academics, managers, and engineers.

The development of a new book about *Higher Education: Progress for Management and Engineering* is essential because:

- It introduces new lines of research in sustainable education in management and engineering fields;
- It contributes to a deeper knowledge in higher education for managers and engineers;
- The theories, strategies, and tools presented and discussed in the diverse chapter contributions will allow researchers, academics, managers, and engineers to take a more strategic role in their organizations.

Many of us, either as academics/researchers or in the different jobs in organizations, found workers with excellent technical and behavioural skills. It is critical for them to know how to do, know how to be, and know how to know. These knowledge, abilities, and competences are obtained from higher education, which is seen as a relevant tool necessary to obtain competitive and dynamic organizations, leading them to the desired success in national and international markets.

Looking to promote research related to these new trends and developments in the field of management and engineering, the present book, divided in seven chapters, covers in Chapter 1 "Integrity in Higher Education: Revisiting Mary Parker Follett to Break New Ground", while Chapter 2 discusses "Digital Transformation in Higher Education Institutions: Key Factors for ERP Projects Leadership". Chapter 3 focuses "Academic Training Contribution to the Development of a Leadership Profile", at the same time, Chapter 4 deals with "Higher Education Policy and Widening Participation: An Overview". Chapter 5 contains information about "The Role of Public Universities in Entrepreneurship in Mexico", while Chapter 6 highlights "Working with Students on Establishing a Student-Oriented Classroom Culture: A Teaching Initiative Designed to Build an Inclusive and Highly Engaging Learning Environment in Online and Face to Face Environments". Finally, Chapter 7 speaks about "Human Resource Management in Education: A Brief Overview".

Higher Education: Progress for Management and Engineering, able to be used by academics, researchers, human resources managers, managers, engineers, and other professionals in related matters, addresses several dimensions of higher education/ sustainable education and its impact on business and the organization competitiveness. Management and engineering in competitive organizations can be deeper improved with high levels of education of its collaborators. Like in other areas of management, it enables managers and engineers with a set of tools, knowledge, and strategies to better manage people, and so entrepreneurs stand to gain from it.

The editors acknowledge their gratitude to CRC Press/Taylor & Francis Group for this opportunity and for their professional support. Finally, we would like to thank to all chapter authors for their interest and availability to work on this project.

Carolina Machado
Braga, Portugal
J. Paulo Davim
Aveiro, Portugal

About the Editors

Carolina Machado received her PhD degree in Management Sciences (Organizational and Policies Management area/Human Resources Management) from the University of Minho in 1999, master's degree in Management (Strategic Human Resource Management) from Technical University of Lisbon in 1994, and a degree in Business Administration from University of Minho in 1989. She has been teaching subjects in Human Resources Management since 1989 at University of Minho, and since 2004 she has been an associate professor, with experience and research interest areas in the fields of Human Resource Management, International Human Resource Management, Human Resource Management in SMEs, Training and Development, Emotional Intelligence, Management Change, Knowledge Management, and Management/HRM in the Digital Age/Business Analytics. She is head of the Human Resources Management Work Group at the School of Economics and Management at University of Minho, coordinator of Advanced Training Courses at the Interdisciplinary Centre of Social Sciences, member of the Interdisciplinary Centre of Social Sciences (CICS.NOVA.UMinho), University of Minho, as well as chief editor of the *International Journal of Applied Management Sciences* and *Engineering (IJAMSE)*, guest editor of journals, books editor, and book series editor, as well as a reviewer of myriad international prestigious journals. In addition, she has also published, both as editor/co-editor and as author/co-author, several books, book chapters, and articles in journals and conferences.

J. Paulo Davim is full professor at the University of Aveiro, Portugal. He is also distinguished as honorary professor in several universities/colleges/institutes in China, India, and Spain. He received his PhD. degree in Mechanical Engineering in 1997, MSc degree in Mechanical Engineering (Materials and Manufacturing Processes) in 1991, Mechanical Engineering degree (five years) in 1986 from the University of Porto (FEUP), the Aggregate title (Full Habilitation) from the University of Coimbra in 2005, and the DSc (higher doctorate) from London Metropolitan University in 2013. He is senior chartered engineer at the Portuguese Institution of Engineers with an MBA and specialist titles in engineering and industrial management as well as in metrology. He is also Eur Ing by FEANI-Brussels and fellow (FIET) of IET-London. He has more than 35 years of teaching and research experience in manufacturing, materials, mechanical, and industrial engineering, with special emphasis on machining and tribology. He has also interest in management, engineering education, and higher education for sustainability. He has guided large numbers of postdoc, PhD, and master's students as well as has coordinated and participated in several financed research projects. He has received several scientific awards and honours. He has worked as an evaluator of projects for ERC-European Research Council and other international research agencies as well as an examiner of PhD thesis for many universities in different countries. He is the editor-in-chief of several international journals, guest editor of journals, books editor, book series editor, and scientific advisory for many international journals and conferences. Presently, he is an editorial board member of 30 international journals and acts as a reviewer for more than 100 prestigious Web of Science Journals. In addition, he has also published (and co-editor) more than 200 books as an editor and as an author (and co-author) more than 15 books, 100 book chapters, and 500 articles in journals and conferences (more than 300 articles in journals indexed in Web of Science core collection/h-index 61+/12,500+ citations, SCOPUS/h-index 66+/15,500+ citations, Google Scholar/h-index 85+/25,500+ citations). He has been listed in world's top 2% scientists by the Stanford University study.

Contributors

Ingrid Yadibel Cuevas Zuñiga
Instituto Politécnico Nacional
Juárez, Mexico

J. Paulo Davim
Department of Mechanical Engineering
University of Aveiro
Campus Santiago
Aveiro, Portugal

G. Gerón-Piñón
Hemispheric & Global Affairs
University of Miami
Coral Gables, Florida

Biginas Konstantinos
School of Business
Hellenic American University
Nashua, New Hampshire

João Fernando Louro
Department of Management, School
    of Economics and Management
University of Minho
Campus Gualtar
Braga, Portugal

Carolina Feliciana Machado
Department of Management, School
    of Economics and Management
University of Minho
Campus Gualtar
and
Interdisciplinary Centre of Social
    Sciences (CICS.NOVA.UMinho)
University of Minho
Braga, Portugal

A. Martins
Graduate School of Business
    and Leadership
University of KwaZulu-Natal
Westville, South Africa

I. Martins
School of Management, IT &
    Governance
University of KwaZulu-Natal
Westville, South Africa

Ericka Molina Ramírez
Instituto Politécnico Nacional
Juárez, Mexico

Daniel Moreira
Department of Management, School
    of Economics and Management
University of Minho
Campus Gualtar
Braga, Portugal

Vlasios Sarantinos
Faculty of Business and Law
University of the West of England
Frenchay Campus
Bristol, United Kingdom

P. Solana-González
Faculty of Economics and Business
    Sciences
University of Cantabria
Santander, Cantabria, Spain

María del Rocio Soto Flores
Instituto Politécnico Nacional
Juárez, Mexico

Sindakis Stavros
School of Business
Hellenic American University
Nashua, New Hampshire

1 Integrity in Higher Education

Revisiting Mary Parker Follett to Break New Ground

A. Martins[1] and I. Martins[2]

[1]Graduate School of Business and Leadership,
University of KwaZulu-Natal, Westville, South Africa

[2]School of Management, IT & Governance, University
of KwaZulu-Natal, Westville, South Africa

CONTENTS

1.1 INTRODUCTION

The financial crises since 2008 spurred the need for higher education institutions (HEIs) to become highly cognisant of the bottom line. However, this has led to the fact that today HEIs, especially universities, produce education that is lower in quality despite being expensive. Universities no longer promote social, public good, but instead, sell a commodity. This chapter gives evidence of how the public HEI model has transitioned from the traditional model to a so-called universal access stage. However, not all universities are able to foment partnerships with industry as the higher-ranked universities do, and this has given rise to the so-called striving universities. The aim of this chapter, a conceptual review, is to reflect on the nexus between university education and entrepreneurial competency in order to bring the important characteristic of integrity back into the scheme of intellectual capital. Indeed, the importance of the ethical dimension emphasizing honesty and disinterestedness contributing to practical wisdom are emphasized.

DOI: 10.1201/9781003021230-1

1.2   THE DEBAZING OF EDUCATION

As a consequence of the growth of neoliberalism and academic capitalism, education has been debased giving rise to lower results in learning, which, in turn, produces a series of cohorts with reduced levels of intellectual capital. The philosophies of Mary Parker Follett (1918) and

> their applications are germane to ensuring the university remains relevant, since they represent a third way to meet the needs of both educators and academicians. She purports that education should develop into a close relationality between student and teacher to bridge theory and practice for the betterment of society [...] The student and teacher are united in a common goal of stimulating student imagination (Follett, 1970). Through properties of relationality, points of agreement can be achieved and a common path forged. If Follettian principles are utilized, fragmented public university educational systems with faculty and administrators engaging in an us vs them struggle may reach consensus on best practices and return to an educational system that focuses on the enhancement of societal, economic and educational norms.
>
> *Pelly and Boje (2020, p. 562)*

In this regard, Follett was of the opinion that educators should improve pedagogy by highlighting the practical aspect or applicability of the theory that is taught. Hence, according to Follett, practical education should be holistic in nature because it should be considered an end as well as education as a means, which is embodied in the imaginative (Follett, 1970). This Follettian framework should be adopted in universities for students to garner analytical skills – considered fundamental for integrative thinking (Follett, 1941).

University graduates make up the largest percentage of those individuals who obtain first-time employment in industry, i.e. 67% (CIPD, 2018). This study on preparedness of young people reinforces the notion that it is vital to have a university degree to enter the job market. However, in the same study, 32% of employers are of the opinion that "graduates are not well prepared" (CIPD, 2018, p. 5). Indeed, a quarter stated that these graduates are "poorly prepared" (CIPD, 2018, p. 5). Hence, the notion is that many universities are disappointing students because they thwart their expectations by not preparing them for life with the necessary skills (Tomlinson, 2008). Ravenscroft and Baker (2020) opine that essential skills are imperative to enhance graduate attributes and safeguard employability. These essential skills entail the following: (i) core skills, (ii) key skills, (iii) functional skills, (iv) skills for life, (v) employability skills, (vi) generic skills, (vii) enterprise skills, (viii) soft skills, (ix) transferable skills and (x) transversal skills. Additionally, university leavers lack motivation, common sense and have a detached attitude (Ravenscroft & Baker, 2020, pp. 10–11). The skills students ought to be acquiring are structural for life as a whole and not only for specific professional skills. Therefore, the university must provide skills to afford students with the know-how in order to be prepared for jobs and passionate in what they do. However, personal development is central to the individual's professional performance and for life as a whole (Grund & Martin, 2012; Suleman & Laranjeiro, 2018).

At this level, the university appears to disappoint graduates enormously, as it does not provide them with the expected practical skills of communication, negotiation, presentation and personal and relational development (Abbasi et al., 2018). Consequently, the university still does not adequately develop individuals for collaborative work, nor is the graduate equipped with the skills to be an enthusiastic, motivated and a committed team member (Balan et al., 2015; Croy & Eva, 2018). This debate about the role of the university developing competencies generates the question of employability (McQuaid & Lindsay, 2005). Moreover, today this notion takes on different forms, compared to the past and its meaning depends on various factors, namely, the individual, the context, relational social capital links, the ease of access to and expansion of information (Brown et al., 2003; Clarke, 2008; Forrier et al., 2015; Gibbs, 2000; Yorke, 2006). This is because employability is much more than simply securing a job; employability involves the sustainability of that job and the degree of pleasure and well-being it provides for the individual (Martins et al., 2014).

Employability is a notion that is linked to students, universities and companies that requires a set of skills which go beyond conventional skills (Hamid et al., 2014; Jackson, 2010; Rae, 2007). However, in light of the current changing society and the concerns surrounding students regarding their integration in the workplace and society, universities are urged to provide a holistic educational path in order for students to be better prepared and supported in their employability trajectory (Wilton, 2008). It is crucial to redefine the cooperative relationship between universities and industry, which, in turn, presupposes the need to also plan a joint curriculum that generates the ability for the individual to be more autonomous, independent, possessing the capability to have the initiative, the creativity, the capacity for conflict resolution, bold leadership skills and overall boldness to embrace challenges (Rae, 2007). Nevertheless, this perspective of cooperation and alliances is still not very common in everyday practice and is viewed as an obstacle which thwarts the development of society. Companies have not yet perceived that they are active role players in the learning process. For this reason, Rae (2007, p. 612) proposes an institutional model of employability-enterprise connectivity leadership, which Rae alludes to as the institutional connectivity of employability and enterprise.

Furthermore, Rae (2007, p. 614) opines that the relationship between employability and organizations requires the development of skills that are interconnected and provides the example of personal, people and task skills, namely:

1. Personal skills:
   - personal organization and time management;
   - self-confidence and self-efficacy;
   - personal budgeting and financial literacy;
   - finding opportunities and taking the initiative to act on opportunities;
   - creative thinking and problem-solving;
   - being able to take decisions and accept risks in conditions of uncertainty;
   - planning, setting goals and persevering to achieve goals; and
   - working independently; taking responsibility for achieving results.

2. People skills:
   - self-presentation and a range of verbal and written communications skills;
   - interpersonal skills of relationship building, negotiation, persuasion and influencing;

- leadership skills in a range of situations;
- team working effectively to achieve results with others; and
- participating in social and industry or professional networks.

3. Task skills:
   - project management;
   - computer literacy and IT skills;
   - numerical, analytical and quantitative skills;
   - being able to apply academic learning in practical settings, including the workplace;
   - being able to adapt and work flexibly in different contexts; and
   - taking responsibility for completing work to quality standards.

In this sense, Rae (2007) further deliberates that work-based learning is essential for the development of personal and applied skills. Furthermore, graduates, in such situations, need to be monitored by companies and accompanied by the university, as well as to be motivated by both institutions in order to reflect on the processes, the situations and the dilemmas. This reflection will enable these graduates to apply in practice the theory they have learned. In this sense, graduates will verify the applicability of the theory and improve it whenever it is adapted to the reality that it faces. A few ways in which this can be achieved are as follows (Rae, 2007, p. 615):

- short-term work experience placement of 6–12 weeks;
- a full academic year work experience placement;
- relevant part-time, casual or vacation work;
- self-employment or freelancing;
- voluntary, community or social enterprise work activity; and
- leadership or organization of student clubs, sports activities or societies.

1.3   UNIVERSITY – FACING THE FUTURE

HEIs have necessarily become aware of the bottom line (Liedtka, 1998) since 2008, as a result of the world financial crisis. This has led to a rising importance of the entrepreneurial university which focused on the so-called academic capitalism that is based on neoliberalism, arising as a consequence of defunding the university and applying market control (Olssen & Peters, 2005; Somers et al., 2018; Touraine, 1992). Furthermore, universities have become, as Slaughter and Leslie (1997, p. 6) termed, the "academic capitalist knowledge-learning regime." Indeed, today HEIs, especially universities produce education that is lower in quality despite being expensive (Slaughter & Rhoades, 2016). Universities no longer promote the social, public good but instead sell a commodity. The public HEI model has moved from the traditional model of free education based on humanistic values committed to the creation of knowledge and its diffusion as well as being directed at the economic and social development of the context in which the organization is based. This focus is not appreciated in a model that is focused on rankings. The latter is the Anglo-Saxon model which relies on excellent research output and reputation of the academic offerings (Ordorika & Lloyd, 2015). Globally, universities are no longer fully funded by their national governments and this has resulted in the change in strategy. Moreover, the university model transitioned from the "elite access stage," [...] to the "massification stage" [and then] to the "universal access stage" (Somers et al., 2018, p. 34) – stage at which universities find themselves in the second decade of the 21st century. Since the end of World War II, moving through to the 1940s and then in the 1980s, the decrease in funding occurred.

Universities in the USA, such as Harvard, have adopted the strategy of disruptive innovation in order to become sustainable, having face-to-face and online programmes at more affordable prices. With the COVID-19 pandemic, this model has further evolved to encompass the hybrid mode of delivery where a few students are on campus, while a larger number are online at the same time. However, not all universities are able to foment partnerships with industry as the higher-ranked universities do, and which are known as the Research Universities. Gonzales et al. (2014, p. 1099) further postulate the "striving universities" which are those HEIs that are in the quest for prestige by paving the way to a dysfunctional context in order to attain the appropriate high-level ranking, at the expense of their faculty members whose aspirations and work-life 'balance' have been adversely affected. For HEIs to be successful, knowing how to identify the main determinants of entrepreneurship, which is the key skill as well as a competency, is thus considered an urgency (Audretsch, 2002; Audretsch et al., 2016). The aim of this chapter is to expound on the 'entrepreneurial university' construct, which is at the core of the triple helix model, and its inherent relationship to the current knowledge society. It is in this context that entrepreneurs claim graduates do not have the necessary skills for the current job market (Bowers-Brown & Harvey, 2004; Cumming, 2010; Heaton et al., 2008). On the other hand, bolder, more enthusiastic and less traditional companies embrace the imperative of organizational learning, which focuses on the notion of learning and unlearning so that individuals are always ready to embrace the rapid skills obsolescence (Argyris & Schön, 1978). This attitude shows an internal predisposition, both from personal and organizational levels, towards teamwork and an unrelenting commitment to personal mastery. In this regard, knowledge creation is considered fundamental for the survival of companies and enables individuals to continuously become upskilled. Therefore, HEIs and industries need to foster partnerships and cooperation between all stakeholders in order to bridge the gap between the skills' deficiencies in the market and the knowledge gained on completion of a HE programme. This is considered vital (Ishengoma & Vaaland, 2016; Lim et al., 2016).

A new university model entails teaching, research and entrepreneurship and is considered the new-Humboldtian 21st century adaptation of the 19th century, which combined education and research, Etzkowitz (2013). The latter postulated the university based on research with the theoretical framework as typified by Wilhelm von Humboldt who amalgamated sciences and humanities and sciences, grounded on viewpoints arising from Fichte and Schleiermacher at the University of Berlin. The university is shifting from simply educating and developing individuals to also sculpting organizations. In Europe, the universities are focusing on preparing the student by introducing Entrepreneurship in the curricula. While in the USA, the focus is on business schools providing programmes for entrepreneurship training. Schumpeter (1951) asserted that entrepreneurship goes beyond the individual to also embrace the organization. While in the late 20th century, universities were not considered profit-making organizations, there has been a fundamental shift because universities require to secure funding and also to support the regional economies.

The paradigm of the entrepreneurial university arose the double helix relationships between university-industry and thereafter university-government that Stanford initiated in the 1900s. These relationships bourgeoned from the co-development of university region. Eventually, these relationships amalgamated into the university-industry-government partnership, as Etzkowitza et al. (2019) corroborate. Etzkowitz and Leydesdorff (2000) perceived that knowledge is created and diffused through a triple helix module which entails a triad grouping of systems, the education system via the university, the economic system via the industry and the political system via the government.

This triple helix was extended to include another helix, namely, public and civil subsystems via a culture and media system, therefore giving rise to the quadruple helix (Carayannis & Campbell, 2009). These additional subsystems support the diffusion of knowledge. In 2010, these authors then added the fifth system to this model – the natural environment, thus giving rise to the quintuple helix (Carayannis & Campbell, 2010). This fifth helix entails elements such as ecology and sustainable development contributing to social innovation and knowledge creation at a societal level. Knowledge in the quintuple helix circulates in a circulate manner (Carayannis et al., 2012). This innovation model provides a path for solving problems and developing sustainability, thus enabling the creation of knowledge.

1.3.1 UNIVERSITY AND TEACHING FOCUSED ON ENHANCING ENTREPRENEURSHIP

The university and its teaching faculty are considered promoters of entrepreneurship because the theory and the abstract that the educator imparts are then applied to the real by the student. Follett (1919, 1941) considers this being the space wherein integration occurs and wherein the importance of the ethical dimension emphasizing honesty and disinterestedness contributing to practical wisdom are emphasized.

An entrepreneur is an agent who motivates, engages others and has an inclusive mindset. Educators are instrumental in mediating change between the constantly evolving world and those students who are about to integrate into the workplace. For this reason, entrepreneurship instruction is a relentless challenge and places special focus on changing the world. Entrepreneurs are change agents that contribute to improving and innovating the economy. However, this spirit needs to be nurtured with complementary skills, many of which do not arise from formal education. Moreover, continuous learning, its structure and the success of innovation processes depend on the entrepreneurial capacity and the relationships that are established between the different institutions, namely, through (i) knowledge and technology, (ii) actors and networks and (iii) institutions (Malerba, 2005).

It is also important to indicate that the Sustainable Development Goals (SDGs), 4 of the 17 SDGs, focus specifically on quality education which should be inclusive and gender equal; this SDG4 is linked to Education 2030 Framework for Action (UNESCO, 2021a, 2021b). Furthermore, the SDG4 encompasses ten targets which include the acquisition of the appropriate set of skills to promote decent work as well as for sustainability and global citizenship. In order to achieve the latter, education needs to encompass different types of entrepreneurship, such as intrapreneurship, social entrepreneurship, green entrepreneurship and digital entrepreneurship. This concern focuses on the need to reconcile the different aspects of teaching entrepreneurship as being complementary to a harmonious development of the individual graduate (Fayolle & Gailly, 2015). Therefore, it is fundamental to generate basic skills by means of teaching. It is in this spirit and taking into account the importance of competence as a fundamental characteristic (Le-Deist & Winterton, 2005) that Chang et al. (2018, p. 860) consider the following competencies to be basic skills for entrepreneurship:

1. *Self-control competency*: Individuals have control over their life, instead of leaving their life in the hands of fate, chance or the ability of others;
2. *Planning or goal-setting competency*: The ability to develop and execute future action plans;
3. Perception and feedback competency: the ability to rationally perceive others and situations and receive and use feedbacks to improve the ability of perception;
4. *Adventure competency*: The ability to rationally and intellectually action under uncertain situations or factors;
5. *Innovation competency*: The ability to use or cite the original ideas of others, develop new situations and ideas;
6. *Decision-making competency*: the ability to respond spontaneously, handle crises, develop solutions plans, select plans and implement them;
7. *Interpersonal competency*: The ability to understand the needs, values and goals of others and timely maintain good relations with others.

For Othman and Muda (2018), emotional intelligence is an entrepreneurial competency that has effects in reducing poverty and unemployment. It is also considered a fundamental element of employability (Goleman, 1999). For this reason, emotional intelligence must be developed throughout the student's academic path and is a core element in the skill set. Moreover, this is part of the soft skills and difficult to conceptualise. However, within this scope, these soft skills can be designated as 'generic skills', 'core skills', 'key skills', 'transferable skills', 'attributes', 'characteristics', 'values', 'competencies', 'qualities', 'personal skills' and 'professional skills' (De La Harpe et al., 2000; Pool & Sewell, 2007; Tymon, 2013). In line with The Pedagogy for Employability Group (2012, pp. 4–5), some of these generic skills are "Imagination/creativity; adaptability/flexibility; willingness to learn; independent working/autonomy; working in a team; ability to manage others; ability to work under pressure; good oral communication; communication in writing for varied purposes/audiences; numeracy; attention to detail; time management; assumption of responsibility and for making decisions; planning, coordinating and organising ability."

On the issue of IT skills, and taking into account circumstances that have arisen as a result of the onset of the COVID-19 context, Liguori and Winkler (2020) substantiate that a new mindset is required when teaching entrepreneurial subjects. Academic faculty members should be going beyond the traditional teaching methods and instead be using new online approaches and methodologies. With the onset of the current COVID-19 pandemic, the face-to-face teaching model has evolved to encompass the various other modes of delivery, which include blended and hybrid. There latter is where a few students are on campus, while a larger number are online at home and at the same time.

1.4   DYNAMIC CAPABILITIES AND KNOWLEDGE CREATION

Indeed, the individual is at the heart of innovation and humans are essential in nurturing and driving dynamic capabilities, thus fortifying dynamic entrepreneurship (Audretsch et al., 2016). Dynamic capabilities are "the firm's ability to integrate, build, and reconfigure internal and external competences to address rapidly changing environments" (Teece et al., 1997, p. 516). Already Penrose (1959) regarded entrepreneurs as part of the dynamic capabilities of the firm. Wernerfelt (1984) and Barney (1991) further expanded on this notion. This further highlights the entrepreneurial approach is inherent in the dynamic capabilities framework. As discussed in this study, organizations need to nurture those abilities that will enable the creation of knowledge. In so doing, this new knowledge, together with methodologies such as that of the socialization, externalization, combination and internalization (SECI) framework, will further enable these organizations to boost their competitive advantage and sustainability by channelling their dynamic capabilities (Nonaka et al., 2016).

Indeed, dynamic capabilities are unleashed by the knowledge-based view of the firm, are creative and adaptive in nature and is the main characteristic that HEIs should aim to be (Grant, 1996; Kozlinska, 2011). The creative nature of dynamic capabilities arises from the interactions among team members. The adaptive capabilities arise from the cognitive capacity of individuals. It is through the *ba*, a shared 'space' and 'time', that leaders should encourage in order to promote and share subjective and contextual knowledge that individuals have and which, in turn, becomes intersubjective knowledge at the organizational level, wherein dynamic capabilities are inherent. "Effort should be made by all stakeholders at all levels in the organization to engage in profound dialogue which is motivating and trust-building and opens up opportunities for knowledge sharing, constructive discord in order to unite all individuals" (Martins et al., 2019, p. 206). Furthermore, the *ba* is energized by distributed leadership which is a feature of the group and not simply the individual (Martins et al., 2017). This type of leadership achieves a substantial part in knowledge creation as well as encouraging the generating and disseminating of knowledge assets. Therefore, distributed leadership stimulates the creative side of dynamic capabilities and it is further nurtured in flexible structures wherein all individual employees in the organization should have a part to play.

A blend of cognitive and noncognitive skills should be aimed at in which self-control, tenacity, sense of purpose and impetus should be balanced with critical thinking, solving problems and memory. "Knowledge is an intangible economic asset and is fundamental for organizational survival and sustainability. Entrepreneurial capacity is nurtured by systemic thinking, which consists of the ability to understand the whole and the inter-relationships that are established among the parts. Therefore, systemic thinking is an ability to express creative thinking" (Martins et al., 2019, p. 202). Moreover, organizational earning goes hand in hand with the nurturing of dynamic capabilities. The ability to understand is supported by reasoning, perception, emotional awareness and conduct on the part of the individual. The level of analysis stemming from the individual contributes to and combines with the deciphering, interpreting and consolidating and fusing that occur at a group level. Thereafter, these different ideas arising from the diverse group members are intermingled into the larger level of the organization as a whole. Subsequent to the 1990s, Carter and Scarbrough (2001) corroborate that both learning and knowledge attained a strategic purpose in organizations as these should advocate learning, nurture and encourage the individual to be a 'learner' (Martins et al., 2017).

1.5 CONCLUSION

In light of the above reflection, extant literature reveals a considerable endeavour is still required to ensure that an entrepreneurial approach is continuously nurtured in the ecosystem, which entails the five helices, the university, industry, government, culture, media and ecology. The symbiosis between these helices, as well as the circular motion that knowledge displays herein, encourages social innovation and knowledge creation and ensures that this ecosystem leads to sustainable development. Indeed, universities will need to focus their attention on providing graduates with the pertinent skill set, knowledge and attitudes required in the post-COVID-19 era wherein workplaces are certainly not the traditional places as they were in the pre-COVID-19 era. Additionally, universities will need to be agile, adaptive and transformative. This will be achieved when universities continue to earnestly collaborate in tandem with industry, government and other stakeholders in order to re-design academic programmes, with various and flexible modes of delivery and access. IT systems will need to be updated to ensure processes that universities use are adapted to these systems. Personalized, flexible and pluralistic systems will also need to be enhanced and be aligned with the diverse nature of the learning needs that modern students have. The Follettian framework should be harnessed in universities so that the learner is equipped with a balance between theory and practice through integrative thinking. This demonstrates that the door opens for further research endeavours.

REFERENCES

Abbasi, F.K., Ali, A., & Bibi, N. (2018). Analysis of skill gap for business graduates: Managerial perspective from banking industry. *Education + Training*, *60*(4), 354–367.

Argyris, C., & Schön, D. (1978). *Organizational learning: A theory of action perspective.* Reading, MA: Addison-Wesley.

Audretsch, D. (2002). Entrepreneurship: A survey of the literature. Centre for economic policy research. Paper prepared for the European Commission, Enterprise Directorate General, 1–70.

Audretsch, D.B., Kuratko, D.F. & Link, A.N. (2016). Dynamic entrepreneurship and technology-based innovation. *Journal of Evolutionary Economics*, *26*, 603–620.

Balan, P., Clark, M., & Restall, G. (2015). Preparing students for flipped or team-based learning methods. *Education + Training*, *57*(6), 639–657.

Barney, J. B. (1991). Firm resources and sustained competitive advantage. *Journal of Management*, *17*(1), 99–120.

Bowers-Brown, T., & Harvey, L., (2004). Are there too many graduates in the UK? A literature review and an analysis of graduate employability. *Industry and Higher Education*, 18(4), 243–254.

Branine, M. (2008). Graduate recruitment and selection in the UK: A study of the recent changes in methods and expectations. *Career Development International, 13*(6), 497–513.

Brown, P., Hesketh, A., & Wiliams, S. (2003). Employability in a knowledge-driven economy. *Journal of Education and Work, 16*(2), 107–126.

Carayannis, E.G., & Campbell, D.F.J. (2009). 'Mode 3' and 'quadruple helix': Toward a 21st century fractal innovation ecosystem. *International Journal of Technology Management, 46*(3/4), 201–234.

Carayannis, E.G., & Campbell, D.F.J. (2010). Triple helix, quadruple helix and quintuple helix and how do knowledge, innovation and the environment relate to each other? A proposed framework for a trans-disciplinary analysis of sustainable development and social ecology. *International Journal of Social Ecology and Sustainable Development, 1*(1), 41–69.

Carayannis, E.G., Barth, T.D., & Campbell, D.F.J. (2012). The quintuple helix innovation model: Global warming as a challenge and driver for innovation. *Journal of Innovation and Entrepreneurship, 1*(2), 1–12.

Chang, J.C., Hsiao, Y., Chen, S.C., & Yu, T.T. (2018). Core entrepreneurial competencies of students in departments of electrical engineering and computer sciences (EECS) in universities. *Education + Training, 60*(7/8), 857–872.

CIPD Workforce Planning Practice (2018). Available at: https://www.cipd.co.uk/Images/workforce-planning-guide_tcm18-42735.pdf [Accessed on 20 January 2021].

Clarke, M. (2008). Understanding and managing employability in changing career contexts. *Employee Relations, 30*(2), 121–141.

Croy, C., & Eva, N. (2018). Student success in teams: Intervention, cohesion and performance. *Education +Training, 60*(9), 1041–1056.

Cumming, J. (2010). Contextualised performance: Reframing the skills debate in research education. *Studies in Higher Education, 35*(4), 405–419.

De La Harpe, B., Radloff, A., & Wyber, J. (2000). Quality and generic (professional) skills. *Quality in Higher Education, 6*(3), 231–243.

Etzkowitz, H. (2013). Anatomy of the entrepreneurial university. *Social Science Information, 52*(3), 486–511.

Etzkowitz, H., & Leydesdorff, L. (2000). The dynamics of innovation: From National Systems and "Mode 2" to a triple helix of university–industry–government relations. *Research Policy, 29*, 109–123.

Etzkowitza, H., Germain-Alamartine, E., Keel, J., Kumar, C., Nelson Smith, K., & Albats, E. (2019). Entrepreneurial university dynamics: Structured ambivalence, relative deprivation and institution-formation in the Stanford innovation system. *Technological Forecasting & Social Change, 141*, 159–171.

Fayolle, A., & Gailly, B. (2015). The impact of entrepreneurship education on entrepreneurial attitudes and intention: Hysteresis and persistence. *Journal of Small Business Management, 53*(1), 75–93.

Follett, M.P. (1918). *The new state: Group organization the solution of popular government.* University Park, PA: Penn State Press.

Follett, M.P. (1919). Community is a process. *The Philosophical Review, 28*(6), 576–588.

Follett, M.P. (1941). In H.C. Metcalf & L.F. Urwick (Eds.), *Dynamic administration: The collected papers of Mary Parker Follett.* New York, NY and London: Harper and Brothers.

Follett, M.P. (1970). The teacher-student relation. *Administrative Science Quarterly, 15*(2), 137–148.

Forrier, A., Verbruggen, M., & De Cuyper, N. (2015). Integrating different notions of employability in a dynamic chain: The relationship between job transitions, movement capital and perceived employability. *Journal of Vocational Behavior, 89*, 56–64.

Gibbs, P.T. (2000). Isn't higher education employability? *Journal of Vocational Education and Training, 52*(4), 559–571.

Goleman, D. (1999). *Working with emotional intelligence.* New York, NY: Bantam Books.

Gonzales, L.D., Martinez, E., & Ordu, C. (2014). Exploring faculty experiences in a striving university through the lens of academic capitalism. *Studies in Higher Education, 39*(7), 1097–1115.

Grant, R. (1996). Prospering in dynamically-competitive environments: Organizational capability as knowledge integration. *Organization Science, 7*(4), 375–387.

Grund, C., & Martin, J. (2012). Determinants of further training – Evidence for Germany. *The International Journal of Human Resource Management, 23*(17), 3536–3558.

Hamid, M.S., Islam, R., & Manaf, N.H. (2014). Employability skills development approaches: An application of the analytic network process. *Asian Academy of Management Journal, 19*(1), 93–111.

Heaton, N., McCracken, M., & Harrison, J. (2008). Graduate recruitment and development. Sector influence on a local market/regional economy. *Education+Training, 50*(4), 276–288.

Ishengoma, E., & Vaaland, T.I. (2016). Can university-industry linkages stimulate student employability? *Education + Training, 58*(1), 18–44.

Jackson, D. (2010). An international profile of industry-relevant competencies and skill gaps in modern graduates. *International Journal of Management Education, 8*(3), 29–58.

Kozlinska, I. (2011). Contemporary approaches to entrepreneurship education. *Journal of Business Management, 4*(1), 205–220.

Le-Deist, F.O., & Winterton, J. (2005). What is competence? *Human Resource Development International, 8*(1), 27–46.

Liedtka, J. (1998). Synergy revisited: How a "screwball buzzword" can be good for the bottom Line. *Business Strategy Review, 8*(2), 45–55.

Liguori, E., & Winkler, C. (2020). From offline to online: Challenges and opportunities for entrepreneurship education following the COVID-19 pandemic. *Entrepreneurship Education and Pedagogy, 3*(4), 1–6.

Lim, Y.M., Lee, T.H., Yap, C.S., & Ling, C.C. (2016). Employability skills, personal qualities, and early employment problems of entry-level auditors: Perspectives from employers, lecturers, auditors, and students. *Journal of Education for Business, 91*(4), 185–192.

Malerba, F. (2005). Sectoral systems of innovation: A framework for linking innovation to the knowledge base, structure and dynamics of sectors. *Economics of Innovation and New Technology, 14*(1), 63–82.

Martins, A., Martins, I., & Pereira, O. (2017). Feedback and feedforward dynamics: Nexus of organizational learning and leadership self-efficacy. In V.C.X. Wang (Ed.), *Encyclopedia of strategic leadership and management* (pp. 207–232). Hershey, PA: IGI Global Publisher of the Business Science Reference.

Martins, A., Martins, I., & Pereira, O. (2019). Entrepreneurship and innovation: The essence of sustainable, smart and inclusive economies. In B. Thomas & L. Murphy (Eds.), *Innovation and social capital in organizational ecosystems* (pp. 195–218). Hershey, PA: IGI Global Publisher of the Business Science Reference.

Martins, A., Martins, I., & Xiao, L. (2014). Employability and talent development in the knowledge economy: What's going on? *International Journal of Social Sustainability in Economic, Social and Cultural Context, 9*(3), 23–36.

McQuaid, R.W., & Lindsay, C. (2005). The concept of employability. *Urban Studies, 42*(2), 197–219.

Nonaka, I., Hirose, A., & Takeda, Y. (2016). 'Meso'-foundations of dynamic capabilities: Team-level synthesis and distributed leadership as the source of dynamic creativity. *Global Strategy Journal, 6*, 168–182.

Olssen, M., & Peters, M.A. (2005). Neoliberalism, higher education and the knowledge economy: From the free market to knowledge capitalism. *Journal of Education Policy, 20*(3), 313–345.

Ordorika, I., & Lloyd, M. (2015). International rankings and the contest for university hegemony. *Journal of Education Policy, 30*(3), 385–405.

Othman, M., & Muda, T. (2018). Emotional intelligence towards entrepreneurial career choice behaviours. *Education & Training, 60*(9), 953–970.

Pelly, R., & Boje, D. (2020). A case for folletian interventions in public universities. *Journal of Applied Research in Higher Education, 12*(4), 561–571.

Penrose, E.T. (1959). *The theory of the growth of the firm*. New York, NY: John Wiley.

Pool, L.D., & Sewell, P. (2007). The key to employability: Developing a practical model of graduate employability. *Education + Training, 49*(4), 277–289.

Rae, R. (2007). Connecting enterprise and graduate employability: Challenges to the higher education culture and curriculum? *Education + Training, 49*(8/9), 605–619.

Ravenscroft, T.M., & Baker, L. (2020). *Towards a universal framework for essential skills*. London: Essential Skills Taskforce.

Schumpeter, J. (1951 [1949]). Economic theory and entrepreneurial history. In *Essays on economic topics*. Port Washington, NY: Kennikat Press Essays.

Slaughter, S., & Leslie, L. (1997). *Academic capitalism: Politics, policies, and the entrepreneurial university*. Baltimore, MD: Johns Hopkins University Press.

Slaughter, S., & Rhoades, G. (2016). State and markets in higher education: Trends in academic capitalism. In M.N. Bastedo, P.G. Altbach, & P.J. Gumport (Eds.), *American higher education in the 21st century: Social, political, and economic challenges*, 4th ed. Baltimore, MD: John Hopkins University Press.

Somers, P., Davis, C., Fry, J., Jasinski, L., & Lee, E. (2018). Academic capitalism and the entrepreneurial university: Some perspectives from the Americas. *Roteiro Joaçaba, 43*(1), 21–42.

Suleman, F., & Laranjeiro, A.M. (2018). The employability skills of graduates and employers' options in Portugal: An explorative study of anticipative and remedial strategies. *Education + Training, 60*(9), 1097–1111.

Teece, D.J., Pisano, G., & Shuen, A.A. (1997). Dynamic capabilities and strategic management. *Strategic Management Journal, 18*, 504–534.

The Pedagogy for Employability Group (2012). *Pedagogy for employability*. Available at: https://www.advance-he.ac.uk/knowledge-hub/pedagogy-employability-2012 [Accessed 21 November 2021].

Tomlinson, M. (2008). The degree is not enough: Students' perceptions of the role of higher education credentials for graduate work and employability. *British Journal of Sociology of Education, 29*(1), 49–61.

Touraine, A. (1992). *Critique de la modernité*. New York, NY: Fayard.

Tymon, A. (2013). The student perspective on employability. *Studies in Higher Education, 38*(6), 841–856.

UNESCO (2021a). *UNESCO – Education 2030: Incheon declaration and framework for action for the implementation of sustainable development goal 4: Ensure inclusive and equitable quality education and promote lifelong learning opportunities for all*. Available at: https://unesdoc.unesco.org/ark:/48223/pf0000245656 [Accessed 24 October 2021].

UNESCO (2021b). *UNESCO sustainable development goals*. Available at: https://en.unesco.org/education2030-sdg4/targets [Accessed 24 October 2021].

Wernerfelt, B. (1984). A resource-based view of the firm. *Strategic Management Journal, 5*(2), 171–180.

Wilton, N. (2008). Business graduates and management jobs: An employability match made in heaven? *Journal of Education and Work, 21*(2), 143–158.

Yorke, M. (2006). *Employability in higher education: What it is – What it is not*. The Higher Education Academy. Available at: https://www.heacademy.ac.uk/knowledge-hub/employability-higher-education-what-it-what-it-not [Accessed 24 October 2021].

2 Digital Transformation in Higher Education Institutions

Key Factors for ERP Project Leadership

G. Gerón-Piñón[1] and P. Solana-González[2]
[1]Hemispheric & Global Affairs, University
of Miami, Coral Gables, Florida, USA
[2]Faculty of Economics and Business Sciences,
University of Cantabria, Santander, Spain

CONTENTS

2.1   INTRODUCTION

The COVID-19 pandemic has forced a new style of living, which is called the new normal. This new normal reflects unfamiliar changes not only in social life, economy, and health but also in educational institutions. It has become clear that leadership is one of the crucial concepts in this pandemic in higher education institutions (HEIs) (Yokuş, 2022). The digitalization has been a bridge of hope for millions of students, and especially traditional roles of educational leaders have changed during this new unfamiliar situation. Social distance principle and students' making use of digital and virtual platforms lead to a change in educational leaders' role of leading, inspiring the digital transformation in higher education (HE) (Marinoni, Van't Land, & Jensen, 2020).

In the past two years, digital transformation has accelerated in HE as institutions have had to manage the effects of the COVID-19 pandemic. Even as they engage more and more in digital transformation (Dx), many institutions still face major barriers to those efforts, the most notable being issues of insufficient cross-institution planning and coordination. Many of these barriers can be managed or overcome by making information technology (IT) processes more versatile, allowing for rapid adaptation to changing needs (Burns, 2021).

In the modern realities of the development of digital economy, universities are facing the need for Dx, the essence of which is based not only on the introduction of digital technologies in the activities of universities but also on the cultural and organizational changes. In this context, an important factor in ensuring the competitiveness of a modern university is the digitalization of its services (Safiullin & Akhmetshin, 2019).

Universities fall behind other sectors, probably due to a lack of effective leadership and changes in culture. This is complemented negatively by an insufficient degree of innovation and financial support (Rodríguez-Abitia & Bribiesca-Correa, 2021).

Leading universities may present a unique challenge because of the organizational complexity of the university, its multiple goals, and its traditional values, as the nature of leadership in HE is ambiguous and contested (Petrov, 2006).

In this context, the role of educational leaders is of considerable importance in HEIs; leaders create learning environments with cultural awareness, serve as collaborators in developing knowledge and engagement, serve as facilitators who promote collaboration, collective responsibility, and an interest in common good (Amey, 2006). Although the very strength of the university system lies in the free and critical thinking, creativity, and autonomy of the people who work in them.

Following the example of large corporations, HEIs are continuously reviewing and improving their management and administration systems. Without any doubt, HEIs are involved in the current technology trends. Technologies like big data analytics, Internet of Things, cloud computing, software as a service (SaaS), cybersecurity, and artificial intelligence are being adopted by educational institutions for improved service delivery (Varma et al., 2021). Consequently, the HE sector has turned to enterprise resource planning (ERP) systems that have shown the promise of enabling them to run their operations more efficiently and at the same time compete better in the academic market (Bhattacharya, 2016).

The ERP systems are often the largest software application adopted by universities with significant amounts allocated to their implementation. However, little research has been conducted about ERPs in a university environment, compared with other environments (ALdayel, Aldayel, & Al-Mudimigh, 2011; Grandon et al., 2020).

The evidence of ERP projects in Latin America is limited, compared with other regions, because just a few HEIs have implemented ERP systems. Since a considerable investment is required, for both the initial acquisition and the ongoing support, the return on investment (ROI) is obtained in the medium or long term, with projects that can take two, three, or more years to be completely implemented. Moreover, although investments in Latin America are not so considerable, they are significant in terms of university's budget.

Jacobson et al. (2007) sustain that Latin America holds the largest compound annual growth rate in ERP spending (21%) at least until 2011. The lack of a more fully developed IT culture might explain the region's lagging international competitiveness. Although ERP-based solutions also appear to be gaining acceptance among Latin American companies, in many cases, such solutions are inadequate, because of the many obstacles preventing ERP from being implemented according to the needs of individual companies.

Therefore, it is highly relevant for HEIs that are considering decisions or are about to start an ERP implementation project to understand the conditions required for this type of project. Many studies on ERP adoption have shown that organizations frequently face several barriers, and the failure rate is high (Althunibat et al., 2019; Bamufleh et al., 2021). This high failure rate is mainly because of its unique characteristic (Chondamrongkul, 2018), where studies have found that 70% of ERP implementations have failed to provide the expected goals (Soliman et al., 2017). For that reason, it is imperative to identify the ERP implementation's best practices in the region, because it is important to minimize the ERP failure in the HE sector (Abugabah & Sanzogni, 2010).

Dx isn't all about technology; there is a human side of the story. Leadership is seen as a key issue in any process of change. However, leaders of professional and knowledge-based institutions, like universities, face special demands (Rocha et al., 2021).

This chapter aims to identify the key factors (KFs) of ERP implementations in HEIs in Latin America with a focus on leadership during Dx, through a compilation of the experience of 23 experts who have participated in projects in 14 countries in the region. It is intended that this research may serve as a reference for institutions that are seeking the implementation of these important systems, and it may serve as a guide for interested stakeholders – financial organisms, owners, investors, managers, responsible for the processes and IT consultants – to start such projects and ensure the understanding of the conditions required that will help them to have a successful project.

The outline of this study is as follows: in the first section, an introduction is presented. In the second part, the concept of Dx in HEIs is stated. The third section presents the ERP systems in the Dx of HEIs. A literature review of KF for ERP Project Leadership is introduced in the fourth section. The fifth section explains in detail the research methodology used. The sixth section shows results and discussion of interviews with experts about the KF, comparisons with other industries, specific characteristics of the Latin American universities, KF in different countries, and finally, the main difficulties that these projects face. The chapter ends with conclusions of this study, its limitations, and future lines of research.

2.2 DIGITAL TRANSFORMATION IN HEIs

Dx is "a series of deep and coordinated culture, workforce, and technology shifts that enable new educational and operating models and transform an institution's operations, strategic directions, and value proposition" (Brooks & McCormack, 2020).

In the context of HE, Dx goes beyond integrating new technologies in the teaching and learning process, with impacts on other fundamental processes such as research, academic and financial management, and course marketing (Burns et al., 2021).

Dx in HEIs requires different factors to achieve this transformation, a process that goes beyond the injection of technology into educational processes, involves the integration of aspects such as skills and updating in the use of technology, the implementation of educational platforms, the adaptation of educational models, as well as the style of leadership (Ramírez, 2021).

In a successful implementation in Baylor University, McCormack et al. (2021) shared some principles when implementing enterprise technology solutions in Dx: more strategic and coordinated approach to aligning systems across functional units; commitment among key stakeholders and institutional leadership to "lock arms" and move forward through the project together; simplification, standardization, transformation; accountable governance; transparency and inclusion; clear communication; measurable results; and Dx in service of the institution's mission.

Supported by an environment that is increasingly immersed in Dx, universities are facing important organizational changes in both tangible and intangible structures. Dx isn't all about technology; there is a human side of the story. Leadership is seen as a key issue in any process of change (Rocha et al., 2021).

The existing management system and infrastructure in universities are often outdated and unable to ensure their competitive and adequate functioning. Hence, there is a need to improve the processes of using the university infrastructure through digital technology (Gafurov et al., 2020).

Dx of processes through IT and systems and ERPs contribute to transforming academic, administrative, and student processes. In addition, the Dx of HEIs must be based on the establishment of strategic plans that contemplate the lines of action that align IT with the institution's objectives. In this sense, the implementation of an ERP provides an efficient response in key areas, the development of electronic administration (including the use of digital certificates and processes with electronic signature), the digitization of the processes that facilitate remote work, and governance of the institution's data.

On the other hand, the virtualization of applications, which are used by faculty and students, the remote deployment of applications, and cloud computing represent very important increases in efficiency and productivity.

The Dx of HEIs also requires the development of specific IT strategic plans that contemplate adequate prioritization, planning, and management of the institution's IT project portfolio, with an adequate endowment of human and financial resources.

2.3 ERP SYSTEMS IN THE DIGITAL TRANSFORMATION OF HEIs

HEIs all over the world strive to increase the efficiency and quality of educational services provided, thereby increasing their competitiveness both within their own country and in the world. Universities are forced to look for new sources of funding and enter into partnerships with private companies; competition is increasingly assuming the characteristics of an open market. All these changes indicate that HE is being gradually transformed into a specific type of industry. HEIs that want to succeed in these changed circumstances must begin to behave more like profit-oriented organizations. They must understand their business processes and, more importantly, know how to enhance and manage their business performance (Serdar, 2010).

HEIs face a disruptive scenario that is established in the new business models, ostensibly transforming the way they evolved over time, actively linking internal and external clients, and increasing their commitment and strengthening their experience in the organization (Serna et al., 2018).

HEIs nowadays focus on leveraging ERP systems as part of their digital strategy since they are facing an imperative need for the implementation of modern technologies to stay competitive and differentiate them as an innovation leader. HE management is challenged with maintaining high-level information systems (Soliman & Karia, 2020).

Enterprise systems are large-scale, real-time, and integrated application-software packages that use the computational, data storage, and data transmission power of modern IT to support processes, information flows, reporting, and business analytics within and between complex organizations (Seddon, Calvert, & Yang, 2010).

The ERP system integrates all facets of business operations that allow information timeliness for the decision-making process, overall efficiency improvement, and achieves competitive advantage (Feldman et al., 2017; Soliman & Karia, 2015).

Pollock and Cornford (2004) suggest that universities share similarities with manufacturing organizations but recognize that universities have specific and unique administrative needs. Traditional ERP systems address basic business administrative functions such as HR (human resource), finance, operations and logistics, and sales and marketing applications. Yet, the HE sector requires unique systems for student administration, course/unit administration, facilities (timetabling) requirements, and other applications, which are not part of traditional ERP. There is an explicit recognition that universities are different: they have components not found in other organizations (students).

However, the differences between universities and organizations are apparent; universities use ERP systems for academic purposes, but organizations use ERP systems for business purposes. Furthermore, ERP is more critical in the HE sector because faculty, staff, and students interact with major educational and administrative activities through ERP (Abugabah & Sanzogni, 2010).

At HEIs, the circumstances that called for moving to an ERP included a redundant, disorganized database structure; inaccurate data; difficulty in reporting and sharing information; dependence on manual processes and human interventions; problems in providing seamless customer service among offices; difficulty complying with reporting requirements; heavy reliance on the computing centre staff; and lack of capacity for process improvements. ERP systems promise to increase operational efficiency, improve customer service, and help enforce an institution's business rules (Powel & Barry, 2005).

There are many reasons that attract universities to implement ERP systems, including global trends, growth in student numbers, a competitive education environment, quality, and performance requirements. These require the HE sector to evolve and replace the existing management and administration systems with ERP systems that provide many management tools and facilities that guarantee the efficiency and accessibility for all users (Abugabah & Sanzogni, 2010).

Academic accreditation standards require the provision of a set of data, documents, reports, and evidence distributed among the various departments of the university (academic and non-academic), which must be provided periodically and annually. ERP systems facilitate the collection of required data from various sources within the university and help to provide the necessary reports and statistics, where the module organizes the necessary processes for quality and accreditation by providing a special database and through linking with the other subsystems at the university (Abdel-Haq, 2020).

In recent years, large companies have invested and developed ERP systems to meet the specific functionalities required for academic administrations that contain their main processes: recruitment, admissions, academic records, courses, registration, class schedules, location management, faculty, accounts receivable, and graduation among others. In Latin America, the most important ERP vendors are Oracle (PeopleSoft), Ellucian (Banner & PowerCampus), and SAP (Campus Solutions). Besides, there are other local vendors such as OCU and ERP university.

ERP systems employ highly integrated business software solutions that have existed for many years. In view of the major digital transformations currently taking place, the role of ERP systems needs to be reconsidered (Asprion, Schneider, & Grimberg, 2018). For example, in respect of the operations, ERP can decide between a multitude of options, e.g. web-based, cloud-based, on premise, mobile apps, and a host of others (Bahssas, AlBar, & Hoque, 2015).

This transformation is fast occurring at the platform tier across the spectrum of core software applications; Morris et al. (2016) call this enhanced ERP portfolio "intelligent ERP", or "i-ERP", and it will run tomorrow's institutions in an increasingly digital world: these intelligent systems will leverage machine learning (ML), cloud deployment, predictive analytics, user experience (UX), collaborative conversations styles (with a mobile-first design) driven by advances in natural language processing (NLP) and ML. Some leading application suppliers have incorporated these technologies: SAP (SAPPHIRE NOW), Salesforce.com (Einstein as "IA for CRM"), IBM (Kenexa Talent Insights powered by Watson), Oracle (smart UX), Microsoft (Axure), and Workday (ML).

At the latest, when the replacement of an outdated ERP system is pending, HEIs need to decide between implementing a more modern version of a classical ERP system or coping with the new approaches related to Dx. enterprise organization. In the past, the classical ERP systems focused on the support of the operational level within an organization and the executive level benefitted from aggregated data and increased transparency due to the harmonized data repository (Ganesh et al., 2014). The next-wave ERP will empower employees in the middle of the organization to efficient operational processes and transparent information flows; an organization will be able to use and enlarge its knowledge base and hence reach productivity gains, thanks to the support of agile communication, collaboration, and interaction (Asprion, Schneider, & Grimberg, 2018).

2.4 KEY FACTORS FOR ERP PROJECT LEADERSHIP

In the HE sector, there is a rigorous need to explore the KFs that lead to a successful implementation of an ERP system. There is relatively little attention and studies that measure ERP success or failure in this sector (Nizamani et al., 2017). ERP implementation practice has proven that making this review provides a good understanding of the elements that contribute to the success (or failure) and also provides a solid base for planning these projects giving a guide to identify the key success factors that need more focus.

Okland (1995) defined KFs as "What the organization must accomplish to achieve the mission by examination and categorisation of the impacts". Verville, Bernadas, and Halingten (2005) claimed that one single KF by itself will not ensure the success of an enterprise system acquisition process, but rather it is a mixture of KFs that will result in the desired outcomes.

Implementing an ERP system is not an inexpensive or risk-free venture. It is therefore worthwhile examining the factors that, to a great extent, determine whether the implementation will be successful (Umble, Haft, & Umble, 2003).

The leadership of HEIs has been placed under increasing scrutiny since the 1980s with the expansion of student numbers, changes in funding for student places, increased marketization and student choice, and continuing globalization of the sector. In this climate of change, HEIs have been required to consider how to develop their leaders and what might be appropriate leadership behaviour to enable adaptation to these new circumstances (Black, 2015).

Kouzes and Posner (2017) studied five practices of exemplary leadership in HEIs: model the way; inspire a shared vision; challenge the process; enable others to act; and encourage the heart.

According to Bryman (2007), effective university leadership requires acting as a role model and having credibility, being considerate, treating staff fairly and with integrity, being trustworthy and having personal integrity; clear sense of direction/ strategic vision, preparing university arrangements to facilitate the direction set, communicating well about the direction the university is going; advancing the university's cause with respect to constituencies internal and external to the university and being proactive in doing so; creating a positive and collegial work atmosphere, allowing the opportunity to participate in key decisions/encouraging open communication; and providing feedback on performance.

Transformative leadership in HE (Astin & Astin, 2000) entails the following characteristics: authenticity, integrity, shared purpose, competence, a learning environment, disagreement with respect; collaboration, division of labour, commitment, and empathy.

The effective leadership and management of universities is a crucial issue for leaders themselves, and university staff. University leadership is fundamentally different from leadership in other contexts and demands additional competencies: academic credibility, experience of university life, research, teaching activities alongside their managerial roles, and people skills, including the ability to communicate and negotiate with others (Spendlove, 2007).

Digital leadership has a strong correlation with transformational leadership; universities need to transform their practices and continue to adapt in order to be effective in the age of Internet. The digital leader is characterized by strategic leadership, business, and digital knowledge (Antonopoulou et al., 2021).

Leadership role-plays address several critical IT issues, including understanding (1) differing objectives of critical project stakeholders; (2) concepts of change management and their importance and process; (3) escalation issues of when and how to do it; and (4) issues arising from client/customer communication (Carton & Richmond, 2018).

Five main aspects of educational leadership are identified as networking, enhancing educational practices, calmness and compassion, analytical and strategical thinking, and transparency, which are supported by educational leadership literature during the pandemic (Chisholm-Burns, Brandon, & Spivey, 2021; Fernandez & Shaw, 2020; Gurr & Drysdale, 2020).

As part of this research, an extensive literature review was conducted of papers related to this topic, looking at those specific articles that analysed this problem in depth and provided best practices in ERP implementations, and also looking at those that referred to implementations in HE. As a result, the articles selected are presented in Table 2.1.

TABLE 2.1
Literature Review on KF of ERP Implementations in HEIs

Author	Objective of the Study
Bamufleh et al. (2021)	This research aims to explore the factors that affect the behavioral adoption and acceptance of an ERP system in the context of HEIs.
Vicedo et al. (2020)	This work presents a bibliometric analysis of the influential authors, institutions, papers, and countries on the field of ERP implementations and their critical success factors based on the Web of Science database with 301 articles belonging to 86 universities and institutions from 48 different countries.
Abuhashish and Al-Tahat (2020)	This research paper describes the scope of the ERP road map recommendation for the Arab Open University Jordan branch administrative services. It provides a high-level overview of the key processes, the major technologies supporting those processes, and the state of readiness for an ERP implementation.
Althunibat et al. (2019)	The main objective of this research is to determine the factors that affect the acceptance of using ERP by Jordanian universities.
Weli (2019)	This research aims to analyse the systems implementation success factors, benefits to users, and satisfaction of users in the implementation of ERP at Atma Jaya Catholic University of Indonesia.
Wanko, Kamdjoug, and Wamba (2019)	This paper seeks to study a case of successful implementation of ERP system in the higher education sector in Cameroon.

(Continued)

TABLE 2.1 *(Continued)*
Literature Review on KF of ERP Implementations in HEIs

Author	Objective of the Study
Chondamrongkul (2018)	This paper presents a case study in a Thai university that implements ERP system, which shows how selected critical success factors are taken into practice throughout implementation process.
Fadelelmoula (2018)	This paper examines empirically the effects of six key critical success factors (CSFs) for the implementation of ERP systems on the comprehensive achievement of the crucial roles of computer-based information systems (CBISs).
Al-Hadi and Al-Shaibany (2017)	The study researches that ERP system will succeed if it is employed with a set of factors such as clear vision and objectives, top management support and commitment, clear business process, information flow and organizational structure, budget size and cost, integrated department and solving the problem of human resources management, project management, training and education, careful change management and effective communication and connection in ERP higher education system.
Nizamani et al. (2017)	This research formulates a conceptual model based on strong background theories to effectively evaluate the success of ERP system implementation.
Ram and Corkindale (2014)	The authors examine the literature on ERP to establish whether the KFs for achieving stages of an ERP project have been empirically shown to be "critical". The authors used a systematic approach to review 627 refereed papers published between 1998 and 2010 on ERP, from which 236 papers related to KFs on ERP were selected for analysis.
ALdayel, Aldayel, and Al-Mudimigh (2011)	This paper explores and analyses the existing literature on ERP implementation and attempts to identify the key success factors for a successful implementation of an ERP in higher education institutions in Saudi Arabia.
Rabaa'i, Bandara, and Gable (2009)	This teaching case illustrates how the contextual factors contribute to the success or failure enterprise systems at Queensland University of Technology (QUT) in Australia.
Rabaa'i (2009)	Through an extensive literature review, the researcher found a large number of articles that provide answers to the question: "What are the key factors for ERP implementation success". These articles were identified through computer searches of many of the outstanding MIS journals.
Umble, Haft, and Umble (2003)	The authors identified the most prominent success factors to a successful implementation after reviewing numerous authors' findings.
Frantz, Southerland, and Johnson (2002)	This study summarizes the survey of higher education chief financial and information officers' perceptions of ERP software implementation best practices. Study participants consisted of 380 chief financial and information officers at 170 institutions accredited by the Southern Association of Colleges and Schools.

Source: Own elaboration.

All this research served as a base for the designing of the experts' interview giving a deep reference regarding the KFs related to leadership that is the aim of this study.

2.5 METHODOLOGY

For this work, qualitative research was carried out based on in-depth interviews with 23 experts. The requisite for selecting these experts was that they had participated in at least one successful implementation of an ERP at an HEI. It was sought that the interviewees would not only be individuals with a management position, for this approach has been ascertained as a deficiency in KF identification (Munro & Wheeler, 1980). Therefore, it was decided to engage with experts who have performed at different levels and roles within an organization (HEI leaders, project managers, implementation team members (ITMs), technical users, consultants, final users, etc.) following the suggestion of Boynton and Zmud (1984), considering that success factors are different for each group.

The selected interviewees were working in the following international companies: Ellucian, SAP, Neoris, Laureate International, Academic Partnerships, Grupo TI, Cornerstone, Riemann Venture, and OCU with an average of 20 years of experience.

Table 2.2 summarizes the interviewees' ERP systems they have implemented and the roles they have performed within the projects in which they were involved.

The interview consisted of five questions related to our topic of inquiry, identifying in the first place HEI-specific KFs in comparison with other industries, and subsequently Latin America's HEI-specific KFs. Below are the questions that served as guidelines for the conducted interviews:

1. Based on your experience, can you tell me which are the factors that you consider key for an ERP system implementation at a HEI to be successful?
2. Do you think there might be any KFs that are specific to HEIs, in comparison with other types of industries?
3. Is there any Latin America-exclusive KF in comparison with other countries?
4. Do you think Latin America's HEIs might have specific or distinctive characteristics to be considered for implementing an ERP system?
5. Which are in your opinion the main difficulties for successful ERP system implementation projects at HEIs? Please sort them by relevance.

The data resulting from the interviews were classified and analysed in detail seeking to keep the content, language, and experience of the interviewees. Preserving the above was found to be critical, as just little has been written in Latin America on this topic. Therefore, it was sought to rescue this knowledge and reflect it in this research, in words of the interviewees themselves focused on the KFs related to leadership of projects for the Dx of HEIs, which is the aim of this research.

TABLE 2.2
Expert Interviewees' Demographic Data Summary

	Implemented Systems				Performed Roles				
Expert	Academic	Financial	HR	Alumni	External PM	Internal PM	Consultant	Implementation Team	Manager
1	•	•	•	•	•	•	•	•	•
2	•	•	•	•	•		•	•	
3	•	•	•	•	•	•	•		
4	•	•						•	•
5	•	•		•	•				
6	•	•					•		
7	•	•					•		•
8	•							•	
9	•	•	•	•	•	•	•	•	•
10	•	•	•	•	•	•	•	•	•
11	•	•					•	•	
12	•	•					•	•	
13	•	•			•	•			•
14	•				•		•		•
15	•	•	•		•			•	
16	•	•	•	•	•	•	•	•	
17	•	•	•				•	•	
18		•	•				•	•	
19	•	•	•				•		
20	•	•	•	•	•	•	•	•	
21		•	•		•			•	
22	•	•	•		•		•		
23	•	•	•	•	•	•			

Source: Own elaboration.

2.6 RESULTS AND DISCUSSION

After an exhaustive classification and analysis process of the information collected during the interviews with experts, 21 KFs were identified as the first result of this research, in relation to questions 1 and 2. Table 2.3 describes these identified KFs.

The results were analysed under three different perspectives:

- *Role analysis*: External project manager (EPM), internal project manager (IMP), consultant (CST), ITM, and manager (MNG) at an HEI.
- *System analysis*: Academic, financial, HRs, and alumni.
- *Country analysis*: Mexico, Brazil, Colombia, Chile, Puerto Rico, Ecuador, Dominican Republic, Venezuela, Peru, Costa Rica, Argentina, Uruguay, Honduras, and Panama.

TABLE 2.3
KFs Identified from the Interviews with Experts

KF	Description
1. Project is aligned to the institution's strategy.	It is very important for an ERP system implementation project to be directly related to the institution's mid- and long-term strategy. There must be a clear focus on the impact of implementing this type of technology, on the institution's reasons to act – the *compelling events* – and the motives behind the institution's decision to implement an ERP – the *drivers*.
2. ERP fulfils the institution's requirements.	The selected ERP must fulfil the daily activities' requirements to which the institution is subject. It is undeniable that, due to local ways of working in each country, or requirements by each Ministry of Education, the ERP is required to have specific functionalities. This is known as "localizations" that are specific developments aimed to address each country's particular requirements.
3. Financial planning (budget during the project, and once it is completed).	For an ERP implementation, institutions are required to make substantial investments in terms of software licensing, maintenance, support, hardware, databases, external consulting, staff, and internal resources. It is important that all these factors are accounted for and budgeted for at the beginning of the project.
4. Clear project expectations (deliverables).	The project expectations must be clearly established, and the key milestones to be achieved as the project develops must be specifically defined as well. It is important for the institution to be well acquainted with the software, also being clear about the benefits it will get from the system, for such clarity will help the institution make achievements during the implementation and meet the established goals.
5. Project planning (scope and timeline).	It is essential to have the required documentation and formality when defining the project scope and timeline before starting the ERP implementation. A project plan must be designed, including its scope, a detailed schedule of the required resources, and timings and costs for the project.
6. Project manager with experience and provided with decision-making capabilities.	A key factor for the assigned project manager is to be experienced in higher education so he/she can preserve the project's leadership and coordination, and for helping both the ERP provider's team and the institution implementation team – *coaching*. It is essential that the project manager is recognized by the institution, and that he/she is provided with decision-making capabilities and the required confidence so he/she can take action. Also, the project manager must have direct communication with the institution's top management, so he/she is able to inform of any situation requiring its attention.
7. Identify who will operate the new system.	It is essential that, prior to the implementation start-up, all roles and individuals who will operate the new system are identified. Those who will operate the new system are properly identified and engaged in the system configuration, during the implementation.
8. Give weekly follow-up to the project advances (progress feedback).	An ERP implementation project at an HEI requires 100% dedicated effort by the implementation team. Therefore, it is essential to make a project progress report on weekly basis. This enables the project leaders to be updated with the achievements reached and provides feedback on the results, for timely acknowledging achieved milestones and goals supporting promptness in case action or a decision is required.

(Continued)

TABLE 2.3 *(Continued)*
KFs Identified from the Interviews with Experts

KF	Description
9. Project communication and change management.	For an ERP system implementation to be successful, it is essential to properly manage change throughout the organization. The project strategic goals must be promoted and communicated from the institution's top management down. The top management must show complete support for these goals and highlight their importance, so they are permeated throughout the institution. A responsible area must be appointed for managing change within the institution, and for informing, training, and sharing the ERP implementation results, as the system is deployed.
10. Top management engagement and support (executive sponsor with decision-making power).	This is undoubtedly one of the most important KFs for an ERP implementation, for the project success mostly depends on the commitment by the institution's top management. It is very important that, from the system selection stages, the top management has a clear understanding of the ERP implementation repercussions. This includes a proper identification of the expected results, and being willing to change, for top management officials are the only ones with authority and power for leading the institution towards achieving the proposed goals. There must be a fully engaged Executive Committee, and such engagement must come right from the institution's President.
11. A committed multidisciplinary team, with experience in the institution and its processes.	Appointing a proper implementation team is substantial for correctly defining and configuring the processes to operate within the ERP. It is essential for the team assigned to the implementation to be deeply acquainted with the institution and its processes, to be experienced in the higher-education-specific practices, and to show an open attitude to innovations and change. All successful implementation projects have in common that their respective implementation teams were 100%, full time engaged in the project.
12. Project governance.	It is important to establish an Executive Committee in charge of providing communication, coordination, and guidance, and making the required decisions during the ERP implementation. According to the experts, it is advisable to have a project manager with the ERP provider, and a project manager from the institution. Both must work together in a coordinated way, supervising the completion of the defined activities towards the achievement of the provided goals. Such structuring enables a governance-specific organization, which enables communication lines to be clearly defined.
13. Standardized critical processes (that can be modelled within the ERP).	It is essential to have all critical processes that are currently functioning at the institution (selection, admission, registration, payment, graduation processes, etc.) properly defined and documented prior to the project implementation start-up. At this stage, the implementation team must reach a consensus on the way the new processes will develop, so they can be modelled within the ERP. This activity is indeed important, for it enables all participants to share the same basis of knowledge for the implementation of procedures, processes, and policies.
14. System modifications because of external requirements.	A policy supporting successful implementation is precisely not making any modification to the system, but only those necessary to address either legal requirements or regulations by the respective Ministry of Education. This measure supports leveraging the system the most and helps to keep it sustainable over time; as there are no modifications to maintain, there won't be any extra costs for the project, and adopting new releases won't be difficult either.

(Continued)

TABLE 2.3 *(Continued)*
KFs Identified from the Interviews with Experts

KF	Description
15. Capacity to innovate and improve the processes.	An ERP provides the chance to innovate and adopt effective business best practices for the education industry. If the institution is willing to innovate and improve its processes, most of the time, the implementation project will succeed. If the institution insists on doing things the way they are used to, it will probably open the door to making modifications to the system which will most likely open, bringing with it delays and high, unbudgeted costs.
16. External specialized consultants support.	It is essential to engage with external consultants specialized in higher-education-specific processes, who can generate trust among the implementation team. This trust creates an environment of certainty that makes it easier to guide the team towards making the changes in the processes and in the institution's way of operating, in cases where these are necessary.
17. Communicate that the ERP project is not just an IT initiative, it's an institutional project.	For an ERP implementation project to be successful, it is essential that the project must not be seen as an IT initiative, but as an institutional project requiring the involvement of the academic and administrative areas. It is suggested that an IT officer as project manager or project leader should not be appointed, but rather someone with prestige among the institution from an academic or administrative area.
18. IT supports.	The IT area becomes an important ally for the ERP provider, as it supports from its role as an implementation team member, the project's technical aspects such as database management, data migration, networks, reporting, system installation, backups, and security management.
19. Efficient decision-making.	During the ERP implementation, the implementation team will need the top management to make decisions in the most efficient way. For instance, decisions must be taken no longer than 72 hours after being identified, as any halt would seriously compromise the project. The required internal structures must be created so decisions needed are swiftly taken during the project execution.
20. Adopt the ERP system standard processes without any modifications.	Adopting the ERP's standard processes is a policy some universities implement from the beginning, aiming to standardize their processes, minimizing costs, and be able to easily adopt new system releases or upgrades.
21. Integration with the current ecosystem.	When selecting the ERP, it is also essential to design a strategy towards the system's proper coexistence and integration with the systems already at the institution, including library systems, learning management systems (LMS), e-mail, portals, business intelligence (BI), and legacy systems. The way the ERP will fit amid the existing ecosystem must be carefully designed.

Source: Own elaboration.

For incorporating the analysis of the three dimensions – Role, ERP system, and Country in which the experts worked – into the identification of the most important 10 KFs for an ERP implementation, the output obtained from the analysis were calculated. Tables 2.4–2.6 describe these results.

TABLE 2.4
KFs for ERP Implementation at an HEI: Analysis by Performed Role

Key Factor	EPM	IPM	CST	ITM	MNG	Σ	#
Top management engagement and support (executive sponsor with decision-making power).	13	6	13	11	7	50	1
A committed multidisciplinary team, with experience in the institution and its processes.	11	4	11	9	4	39	2
Project communication and change management.	7	3	11	10	5	36	3
Project is aligned to the institution's strategy.	7	3	8	7	4	29	4
Clear project expectations (deliverables).	7	–	9	8	3	27	5
Standardized critical processes (that can be modelled within the ERP).	5	3	8	5	3	24	6
ERP fulfils the institution's requirements.	4	2	7	6	3	22	7
Project planning (scope and timeline).	4	3	3	6	3	19	8
Financial planning (budget during the project, and once it is completed).	4	2	3	5	3	17	9
Capacity to innovate and improve the processes.	2	1	6	3	3	15	10

Source: Own elaboration.

Overall, the results of the analyses carried out indicate that the most important KF for an ERP implementation project is the "top management engagement and support (executive sponsor with decision-making powers)". The 23 experts indicated this, sharing the experiences from those institutions at which the highest authority – the institution's President or owner – led the project and supported it on a permanent basis; tracked the project's progress; generated decision-making mechanisms; and actively participated in the project's internal and external promotion. It is essential that the top management engages in the project and gets actively involved in it. It is not enough only to allocate the resources and appoint an implementation team in charge of the project, but they must also understand the project, since this type of implementations are, for institutions engaging into them, equivalent to having their processes "peeled to the bone" (quoting one of the experts). Therefore, a strong leadership and monitoring are essential for achieving the project goals. The following most important KFs are secondary to an adequate leadership, highlighting the need for the sponsor to understand in detail the repercussions of implementing this technology, and the benefits resulting from it, which enables the sponsor to lead the institution through a long implementation and helps it assimilate the new processes.

TABLE 2.5

KFs for ERP Implementation at an HEI: Analysis by System Implemented

Key Factor	Academic System	Financial System	Human Resources System	Alumni System	Σ	#
Top management engagement and support (executive sponsor with decision-making power).	19	18	12	9	58	1
A committed multidisciplinary team, with experience in the institution and its processes.	15	15	11	7	48	2
Project communication and change management.	15	13	8	4	40	3
Project is aligned to the institution's strategy.	13	12	7	5	37	4
Clear project expectations (deliverables).	12	11	8	3	34	5
Standardized critical processes (that can be modelled within the ERP).	9	7	5	3	24	6
ERP fulfils the institution's requirements.	8	7	4	1	20	7
Financial planning (budget during the project, and once it is completed).	7	6	3	2	18	8
Project planning (scope and timeline).	6	6	3	2	17	9
Give weekly follow-up to the project advances (progress feedback).	5	4	3	2	14	10

Source: Own elaboration.

Weli (2019) found out that strong support from leaders, favourable project management, and appropriate change management may anticipate user resistance and for instance support successful implementations. In addition, according to Burns et al. (2021), IT has been put on top of the strategic agenda, paving the way for Dx initiatives reinforcing that top management support is the most important KF for successful ERP strategic projects.

The performance of organizations, including universities, depends on the alignment of their educational strategy with the evolution of technology (Wanko, Kamdjoug, & Wamba, 2019).

Appointing a proper implementation team is substantial for correctly defining and configuring the processes to operate within the ERP. It is essential for the team assigned to the implementation to be deeply acquainted with the institution and its processes, to be experienced in the HE-specific practices, and to show an open attitude to innovations and change. All successful implementation projects have in common that their respective implementation teams were 100%, full time engaged in the project.

It is very important for an ERP implementation project to be directly related to the institution's mid- and long-term strategy. There must be a clear focus on the impact

TABLE 2.6
KFs for ERP Implementation at an HEI: Analysis by Country of Operation

Key Factor	Mexico	Brazil	Colombia	Chile	Puerto Rico	Ecuador	R. Dominican	Venezuela	Peru	Costa Rica	Argentina	Uruguay	Honduras	Panama	Σ	#
	11	6	15	13	9	7	7	8	9	7	5	1	1	1	100	1
Top management engagement and support (executive sponsor with decision-making power).	9	3	11	9	8	5	6	5	8	6	2	–	1	1	74	2
A committed multidisciplinary team, with experience in the institution and its processes.	8	5	9	9	5	4	6	4	5	4	3	1	1	–	64	3
Project is aligned to the institution's strategy.	9	3	9	5	6	5	2	4	6	4	2	1	–	1	57	4
Project communication and change management.	7	2	7	5	6	4	4	4	6	6	2	–	–	1	54	5
Standardized critical processes (that can be modelled within the ERP).	9	3	6	7	5	3	4	2	4	4	2	–	1	–	50	6
Clear project expectations (deliverables).	9	2	4	4	3	3	1	2	2	2	1	–	–	–	33	7
ERP fulfils the institution's requirements.	4	3	4	5	1	1	2	2	1	1	3	1	–	–	28	8
Financial planning (budget during the project, and once it is completed).	5	2	3	5	3	1	1	1	2	1	2	1	–	–	27	9
Project planning (scope and timeline).	5	2	3	5	3	1	1	1	2	1	2	1	–	–	25	10
Give weekly follow-up to the project advances (progress feedback).	5	–	3	3	3	3	1	1	3	2	1	–	–	–		

Source: Own elaboration.

of implementing this type of technology, on the institution's reasons to act – the compelling events – and the motives behind the institution's decision to implement an ERP. Creating clear organizational goals sets your transformation up for success. Once goals and objectives are well defined, the organization has a clear direction forward towards building digital environments that are agile, flexible, and aligned with institutional strategy (Burns, 2021).

For an ERP system implementation to be successful, it is essential to properly manage change throughout the organization. The project strategic goals must be promoted and communicated. A responsible area must be appointed for managing change within the institution, and for informing, training, and sharing the ERP implementation results, as the system is deployed.

It is essential to have all critical processes that are currently functioning at the institution properly defined and documented prior to the project implementation start-up. At this stage, the implementation team must reach a consensus on the way the new processes will develop, so they can be modelled within the ERP. This activity is important for it enables all participants to share the same basis of knowledge for the implementation of procedures, processes, and policies.

The project expectations must be clearly established, and the key milestones to be achieved as the project develops must be specifically defined as well. It is important for the institution to be clear about the benefits it will get from the system, for such clarity will help the institution make achievements during the implementation and meet the established goals.

The selected ERP must fulfil the daily activities' requirements to which the institution is subject. It is also important for a successful ERP implementation to make substantial investments in terms of hardware, software licensing, databases, external consulting, support, staff, and internal resources. It is important that all these factors are accounted for and budgeted for at the beginning of the project. Defining the scope, timeline, and a process to give weekly follow-up to the project and report progress is KF for a timely and rigorous project administration. In addition, the team needs to be able to innovate and improve the process while performing the ERP implementation always with the support of the leaders.

Continuing with the results of the interviews, regarding the HEI-specific KFs in comparison with other types of industries (question 2) below is a summary of the interviewer's insights:

- *Decisions are made by consensus; therefore, the decision-making process is slow as the academic cycle at the university*: Faculty play a key role in decision-making processes at HEIs; in most cases, there is a Faculty Senate which is a vehicle by which the faculty is authorized to share in planning and governance of the university. The Senate plays an important role in dealing with academic policies, proposals for developing new programmes, compensation and the annual budgets, reviews of chairs and deans, and countless other matters of collegial concern and interest. It is important that leaders partner not only with faculty in the implementation of digital transformation processes but also with key administrators, board of trustees, students' organizations, etc.

- *There is a politicized environment at the university*: Politics has always been embedded in HE. That's because critical thinking, ethics, science, research, free speech, and social justice are foundational elements for colleges and universities around the world. Its imperative is to recognize the nature of the political relations between key stakeholders within the university to make them part of the project during the communications, change management, and the decision-making processes.
- *There are no standardized processes; each university operates differently*: Every university has its own identify, differentiation and competitive advantage. In Europe and United States, standardization is a central feature emphasized by International Organizations like the Organization for Economic Cooperation and Development, but in Latin America, each university operates according to the reference model, American or Spanish, which makes it difficult when implementing ERPs.
- *The staff working on the project is in very good standing and have a high academic level*: In HEIs, having a high academic level is valued, expected, and compensated. This is different from other industries where the experience is more appreciated.

It is very important for any professionals that have not previously worked amid the HE environment to recognize these features making HEIs unique, in such a way that understanding these features and bearing them in mind enable the professionals to take the right steps for addressing them. The discovery about decision-making by consensus, which was mentioned by 13 experts, is reasserted by the study by Heiskanen, Newman and Simila (2000), which detailed analysis of the use of software packages, leading to the conclusion that standard systems in use at the HE industry are not adequate for universities as organizations, for these are unique especially concerning their decision-making processes. It would appear that universities are essentially different than business organizations concerning their decision-making process. In consequence, the information system development standard created for businesses might not be adequate for HEIs.

Therefore, as the KF analysis concluded, establishing an infrastructure and mechanisms enabling "efficient decision-making" is essential for a successful ERP implementation, as it is known that several projects stagnated after a pending decision was left to the next scheduled Academic Committee meeting.

To question 3, the experts answered indicating the Latin American universities-specific characteristics, which are as follows:

- *There is a lack of execution discipline; environment is informal*: The work environment is informal characterized by processes around the attention to students, who are the main clients of the universities. Successful leadership requires understanding, focused, and be disciplined to work at it, until it becomes a natural way of working. In HE means, adopting best practices of other business areas such as finance, HRs, strategic planning, and formalizing those processes is done.

- *Concerning the people's participation model, there is little initiative*: As previously shared, HEIs are not very disposed to change, so people who participate in the ERP implementation systems will have little initiative and will be waiting to participate according to what is indicated.

Also critical for a successful project is understanding, the HEI context in Latin America, where standard processes are few in comparison with other regions; and Ministries of Education in each Latin American country continuously change regulations, making it difficult to implement an ERP, which is based on best practices and promotes them. Therefore, the "standardized critical processes (that can be modelled within the ERP)", KF this study reveals, were not mentioned in the reviewed literature, for it is a Latin America-specific characteristic. In consequence, each university thinks it is unique for it designed its own way to operate, most frequently without any reference at all; there is no professionalization the way it is at developed countries in which there are a higher specialization. In Latin America, this subject is just being recognized, and it is acknowledged that the region is many years behind in comparison with world powers.

Question 4 was aimed to go deeper into the differential aspects exclusive to Latin America. Results accordingly with the interviewed experts are:

- *Lack of professional management*: The professionalization of HE in Latin America is limited but recent efforts are emerging, for example the National Association of Universities and Higher Education Institutions (ANUIES) IT annual meeting, a space aimed at IT directors and managers of HEIs in Ibero-America with specialized workshops for training and updating on IT issues to potentiate emerging technologies.
- *The relationship with the people is very important; you must win them over; it is necessary to make them engage*: In Latin America, the collective spirit of the workplace is manifested in several ways, the importance of personal relationships and loyalty to the group. For instance, it is very important to engage with the people involved in the project.
- *The political and socio-economic environment within the university does not support the adoption of new technologies and ways to operate*: Faculty does not support the adoption of new technologies unless they are used directly in the classroom. Is for this reason that for ERP implementations is important to identify those professors who are open to learning and adapting other technologies that will support the academic processes, for example the implementation of a new online gradebook with early alerts integrated with the learning management system that will support the predictive model of students successfully finishing a course.

This is important for professionals experienced in this project to understand, when starting a project at an HEI in Latin America, which the institution will most likely lack any reference at all about what an ERP project involves. Therefore, the institution will need to be supported, from making it understand what an ERP technology implementation project is, to jointly identifying how the ERP project will impact the institution's strategic goals. Skipping this step will make embracing the new system very difficult for the institution, as ERP implementation projects impact both academic and administrative areas.

Other elements the experts commented on the most were cultural aspects, essentially highlighting the importance of relationships between people. Since universities' life develops among a political environment, it is important to get closer to all the people that would be somehow affected by an ERP project and convince them; you must win them over; both the project and its results must be explained to them, for the opposition of those even at lower hierarchical levels can seriously affect the project. The experts' recommendation is to identify those against the project right at its first stages, when the processes are being reviewed and designed. Once the individuals against the project are identified, special activities must be carried out with them as part of the project's risks identification, so actions can be made on the issue.

The last question was aimed to validate the entire survey, asking the experts to indicate the most important difficulties hindering an ERP implementation process. Table 2.7 shows the results from the expert opinion collected data, sorted by relevance.

The first challenge that ERP implementation projects face is resistance to change. The experts commented that the fear and misunderstanding within the institution when an ERP implementation underway leads the institution's staff to object to the project and block its activities. On this concern, the human factor is essential for pulling a project through an HEI, since these projects require the engagement of all levels within the organization as previously commented, for everyone will be impacted by the new processes.

An effective transformational leader aims to motivate followers, seek to meet their highest needs, and gain their full commitment. This is associated with charisma and vision and is emerging as the most appropriate model, as it focuses on issues of change and transformation (Bryant, 2003). These characteristics are important for a leader to face the main difficulties when implementing ERP systems: resistance to change, continual change or priorities and expectations, and lack of top management and institutional commitment.

TABLE 2.7

Difficulties for ERP Successful Implementation Projects at HEIs

Challenges to be Overcome by ERP Project Leaders	Number of Experts
Resistance to change.	10
Continual change of priorities and expectations.	8
Lack of a single team properly acquainted with the processes.	7
Lack of top management and institutional commitment.	4
The ERP implementation is not positioned as an institutional project led by the President.	4
Lack of a well-positioned project leader.	3

Source: Own elaboration.

In a recent study on leadership for effective governance in HEIs (Hemakumar, 2021), the following personalities were identified as necessary to become successful institution leaders:

- Leaders have a vision and a plan.
- Leaders create collaborative and inclusive learning environments.
- Leaders are passionate about their works.
- Leaders lead by example (role model).
- Leaders understand the importance of community building.
- Leaders encourage risk-taking.
- Leaders are lifelong learners.

These findings are important for leaders who are interested in successfully implementing ERP systems in their institutions in a complex and challenging context of Dx.

2.7 CONCLUSIONS

This study was conducted focusing on a subject where there is such little evidence in the existing literature on what the KFs of ERP implementations in HEIs are. The relevance of this research is its focus in Latin America, a young region in which just a few HEIs have implemented ERPs. Therefore, it was very attractive and enriching to produce this chapter that may serve as a reference for the institutions that are planning to implement these projects and help to make them successful.

An important finding that was recognized by the experts is that the implementation team has to be 100% focused on the project. Projects that have been successful are those where the implementation team was focused and dedicated to the project, and during this time, the top management supported them with additional personnel responsible for assisting them in their daily activities, also giving them the confidence that their job was safe, and in many cases, people involved in the implementation became key persons at the institution, so in addition to recognition by the project, they had growth and promotion opportunities inside the institution. It is not enough to involve key people; it is essential to allocate them full time.

As was demonstrated in the analysis, the identified difficulties can be mitigated if the HEIs commit with the KFs identified in this research since eight of the ten most important KFs act as moderators.

It is important to point out that 18 of the 23 experts that participated in this study had performed a top management position in an HEI, internal project management or had been a member of the university implementation team. Therefore, researchers evaluated the findings in the interviews and concluded that the information gathered was sufficient and contained the opinions of the people who could meet the HEIs, so it was decided to not extend this study to surveys in universities.

The effective leadership and management of universities is a crucial issue for leaders themselves, and for university staff. Leadership in Dx means nothing but creating an environment for people to grow as leaders; interacting effectively and clearly; creating an environment for the achievement of goals; positioning, directing, managing conflict, and shaping norms; and maintaining presence.

For future research, it is recommended to make this study by university type, because as was noted previously, in Latin America, the percentage of private HEIs is large and is different in their operation, financial capacity, and management compared to public HEIs.

REFERENCES

Abdel-Haq, M. S. (2020). Conceptual framework for developing an ERP module for quality management and academic accreditation at higher education institutions: The case of Saudi Arabia. *International Journal of Advanced Computer Science and Applications*, *11*(2), 144–152.

Abugabah, A., & Sanzogni, L. (2010). Enterprise resource planning (ERP) system in higher education: A literature review and implications. *International Journal of Human and Social Sciences*, *5*(6), 395–399.

Abuhashish, F., & Al-Tahat, K. (2020). Readiness research for implementing ERP systems: A case study. *International Journal of Business Information Systems*, *35*(2), 167–184.

ALdayel, A. I., Aldayel, M. S., & Al-Mudimigh, A. S. (2011). The critical success factors of ERP implementation in higher education in Saudi Arabia: A case study. *Journal of Information Technology and Economic Development*, *2*(2), 1–16.

Al-Hadi, M. A., & Al-Shaibany, N. A. (2017). Critical success factors (CSFs) of ERP in higher education institutions. *International Journal of Advanced Research in Computer Science and Software Engineering*, *7*(4), 92–95.

Althunibat, A., Zahrawi, A. A., Tamimi, A. A., & Altarawneh, F. H. (2019). Measuring the acceptance of using enterprise resource planning (ERP) system in private Jordanian universities using tam model. *International Journal of Information and Education Technology*, *9*(7), 502–505. doi: 10.18178/ijiet.2019.9.7.1254

Amey, M. J. (2006). Leadership in higher education. *Change: The Magazine of Higher Learning*, *38*(6), 55–58. doi: 10.3200/CHNG.38.6.55-58

Antonopoulou, H., Halkiopoulos, C., Barlou, O., & Beligiannis, G. (2021). Digital leader and transformational leadership in higher education. *Proceedings of INTED2021 Conference.*

Asprion, P. M., Schneider, B., & Grimberg, F. (2018). ERP systems towards digital transformation. In: Dornberger R. (ed) *Business Information Systems and Technology 4.0. Studies in Systems, Decision and Control*, vol 141. Springer, Cham. doi: 10.1007/978-3-319-74322-6_2

Astin, A. W., & Astin, H. S. (2000). *Leadership Reconsidered: Engaging Higher Education in Social Change*. Michigan, United States: ERIC.

Bahssas, D. M., AlBar, A. M., & Hoque, R. (2015). Enterprise resource planning (ERP) systems: Design, trends and deployment. *The International Technology Management Review*, *5*(2), 72–81.

Bamufleh, D., Almalki, M. A., Almohammadi, R., & Alharbi, E. (2021). User acceptance of enterprise resource planning (ERP) systems in higher education institutions: A conceptual model. *International Journal of Enterprise Information Systems*, *17*(1), 144–163. doi: 10.4018/IJEIS.20210101.oa1

Bhattacharya, P. (2016). Strategizing and innovating with enterprise systems: The case of a public university. *Journal of Cases on Information Technology*, *18*(2), 1–15. doi: 10.4018/JCIT.2016040101

Black, S. A. (2015). Qualities of effective leadership in higher education. *Open Journal of Leadership*, *4*(2), 54–66. doi: 10.4236/ojl.2015.42006

Boynton, A. C., & Zmud, R. W. (1984). An assessment of critical success factors. *Sloan Management Review*, *25*(4), 17–27.

Brooks, D. C., & McCormack, M. (2020). *Driving Digital Transformation in Higher Education*. Colorado, United States: EDUCAUSE.

Bryant, S. E. (2003). The role of transformational and transactional leadership in creating, sharing and exploiting organizational knowledge. *Journal of Leadership & Organizational Studies*, *9*(4), 32–44. doi: 10.1177/107179190300900403

Bryman, A. (2007). Effective leadership in higher education: A literature review. *Studies in Higher Education*, *32*(6), 693–710. doi: 10.1080/03075070701685114

Burns, S. (2021). *Flexibility, Agility, and the Three Dx Shifts: Culture, Workforce, and Technology*. Colorado, United States: EDUCAUSE.

Burns, S., Rodríguez, J. R., Gil-Rodríguez, E. P., & Caballé, D. (2021). *Dx through Workforce and Governance Shifts at Universitat Oberta de Catalunya: An EDUCAUSE Research Case Study*. Colorado, United States: EDUCAUSE.

Carton, R., & Richmond, W. (2018). IT leadership and ERP: A challenging day for a new leader. *Journal of Information Technology Teaching Cases*, *8*(2), 209–216. doi: 10.1057/s41266-018-0039-5

Chisholm-Burns, M. A., Brandon, H. H., & Spivey, C. A. (2021). Leadership lessons from administrators, faculty, and students during the COVID-19 pandemic. *Currents in Pharmacy Teaching and Learning*, *13*(10), 1306–1311. doi: 10.1016/j.cptl.2021.07.001

Chondamrongkul, N. (2018). ERP implementation in university: A case study in Thailand. *International Journal of Business Information Systems*, *27*(2), 177–192.

Fadelelmoula, A. A. (2018). The effects of the critical success factors for ERP implementation on the comprehensive achievement of the crucial roles of information systems in the higher education sector. *Interdisciplinary Journal of Information, Knowledge, and Management*, *13*, 21–44. doi: 10.28945/3942

Feldman, G., Shah, H., Chapman, C., Pärn, E. A., & Edwards, D. J. (2017). A systematic approach for enterprise systems upgrade decision-making: Outlining the decision processes. *Journal of Engineering, Design and Technology*, *15*(6), 778–802. doi: 10.1108/JEDT-08-2017-0076

Fernandez, A. A., & Shaw, G. P. (2020). Academic leadership in a time of crisis: The coronavirus and COVID-19. *Journal of Leadership Studies*, *14*(1), 39–45. doi: 10.1002/jls.21684

Frantz, P. S., Southerland, A. R., & Johnson, J. T. (2002). ERP software implementation best practices. *Educause Quarterly*, *25*(4), 38–45.

Gafurov, I. R., Safiullin, M. R., Akhmetshin, E. M., Gapsalamov, A. R., & Vasilev, V. L. (2020). Change of the higher education paradigm in the context of digital transformation: From resource management to access control. *International Journal of Higher Education*, *9*(3), 71–85. doi: 10.5430/ijhe.v9n3p71

Ganesh, K., Mohapatra, S., Anbuudayasankar, S. P., & Sivakumar, P. (2014). *Enterprise Resource Planning: Fundamentals of Design and Implementation*. Berlin/Heidelberg, Germany: Springer.

Grandon, E. E., Magal, S. R., Pinzon, B. H. D., & Contreras, K. R. (2020). Validation of an ERP system acceptance model among Latin American students: A longitudinal study in Chile and Colombia. *Iberian Conference on Information Systems and Technologies, CISTI*, art. no. 9140891. doi: 10.23919/CISTI49556.2020.9140891

Gurr, D., & Drysdale, L. (2020). Leadership for challenging times. *International Studies in Educational Administration*, *48*(1), 24–30.

Heiskanen, A., Newman, M., & Simila, J. (2000). The social dynamics of software development. *Accounting, Management & Information Technologies*, *10*(1), 1–32. doi: 10.1016/S0959-8022(99)00013-2

Hemakumar, G. (2021). Study on academic leadership for effective governance in HEIs. *International Journal of Arts, Science and Humanities*, *8*(1), 218–224. doi: 10.34293/ash.v8iS1-Feb.3955

Jacobson, S., Shepherd, J., D'Aquila, M., & Carter, K. (2007). *The ERP Market Sizing Report, 2006–2011*. Düsseldorf, Germany: AMR Research.

Kouzes, J. M., & Posner, B. Z. (2017). *The Leadership Challenge Workbook*. New Jersey, United States: John Wiley & Sons.

Marinoni, G., Van't Land, H., & Jensen, T. (2020). *The impact of Covid-19 on higher education around the world*. IAU Global Survey Report. International Association of Universities.

McCormack, M., Dalton, B., Gochis, C., & Allen, J. (2021). *Enterprise Technology As a Catalyst for Digital Transformation at Baylor: An EDUCAUSE Research Case Study*. Colorado, United States: EDUCAUSE.

Morris, H. D., Mahowald, R. P., Jimenez, D. Z., Stratis, A., Rizza, M. N., Hayward, D., & Motai, Y. (2016). *i-ERP (Intelligent ERP): The New Backbone for Digital Transformation*. Massachusetts, United States: IDC Analyze the Future.

Munro, M. C., & Wheeler, B. R. (1980). Planning, critical success factors, and management's information requirements. *MIS Quaterly, 4*(4), 27–38.

Nizamani, S., Khoumbati, K., Ismaili, I. A., Nizamani, S., Nizamani, S., & Basir, N. (2017). Testing and validating the ERP success evaluation model for higher education institutes of Pakistan. *International Journal of Business Information Systems, 25*(2), 165–191.

Okland, J. S. (1995). *Total Quality Management – Text with Cases*. Oxford, United Kingdom: Butterworth-Heinemann.

Petrov, G. (2006). The leadership foundation research on collective leadership in higher education. *Leadership Matters, 7*(11), 11.

Pollock, N., & Cornford, J. (2004). ERP systems and the university as a "unique" organization. *Information Technology and People, 17*(1), 31–52. doi: 10.1108/09593840410522161

Powel, W. D., & Barry, J. (2005). An ERP post-implementation review: Planning for the future by looking back. *Educause Quaterly, 28*(3), 40–46.

Rabaa'i, A. A. (2009). Identifying critical success factors of ERP Systems at the higher education sector. *Third International Symposium on Innovation in Information & Communication Technology, Amman, Jordan*.

Rabaa'i, A. A., Bandara, W., & Gable, G. G. (2009). ERP systems in the higher education sector: A descriptive case study. *Proceedings of the 20th Australasian Conference on Information Systems, Melbourne*.

Ram, J., & Corkindale, D. (2014). How "critical" are the critical success factors (CSFs)?: Examining the role of CSFs for ERP. *Business Process Management Journal, 20*(1), 151–174. doi: 10.1108/BPMJ-11-2012-0127

Ramírez, M. R. (2021). Digital transformation in the universities: Process in the time of covid 19. *Revista Iberica de Sistemas e Tecnologias de Informacao, E42*, 573–582.

Rocha, A., Gonçalves, M. J. A., da Silva, A. F., Teixeira, S., & Silva, R. (2021). Leadership challenges in the context of university 4.0. A thematic synthesis literature review. *Computational and Mathematical Organization Theory*, 1–33. doi: 10.1007/s10588-021-09325-0

Rodríguez-Abitia, G., & Bribiesca-Correa, G. (2021). Assessing digital transformation in universities. *Future Internet, 13*(2), 52–68. doi: 10.3390/fi13020052

Safiullin, M. R., & Akhmetshin, E. M. (2019). Digital transformation of a university as a factor of ensuring its competitiveness. *International Journal of Engineering and Advanced Technology, 9*(1), 7387–7390. doi: 10.35940/ijeat.A3097.109119

Seddon, P. B., Calvert, C., & Yang, S. (2010). A multi-project model of key factors affecting organizational benefits from enterprise systems. *MIS Quarterly, 34*(2), 305–328.

Serdar, A. M. (2010). Performance management and key performance indicators for higher education institutions in Serbia. *Perspectives of Innovations, Economics and Business, 6*(3), 116–119.

Serna, M. D. A., Branch, J. W., Benavides, L. M. C., & Burgos, D. (2018). Un modelo conceptual de transformación digital. Openergy y el caso de la Universidad Nacional de Colombia. *Education in the Knowledge Society, 19*(4), 95–107. doi: 10.14201/eks201819495107

Soliman, M., & Karia, N. (2015). Enterprise resource planning (ERP) system as an innovative technology in higher education context in Egypt. *International Journal of Computing Academic Research, 5*(4), 265–269.

Soliman, M., & Karia, N. (2020). Explaining the competitive advantage of enterprise resource planning adoption: Insights Egyptian higher education institutions. *Journal of Information Technology Management, 12*(4), 1–21. doi: 10.22059/JITM.2020.292788.2424

Soliman, M., Karia, N., Moeinzadeh, S., Fauzi, F. B. A., & Islam, M. S. (2017). Towards an understanding of the behavior intention and benefits to use enterprise resource planning systems among higher education institutions' end-users in Egypt: The role of readiness for change. *Proceedings of the 12th Asian Academy of Management International Conference*, Penang, Malaysia.

Spendlove, M. (2007). Competencies for effective leadership in higher education. *International Journal of Educational Management, 21*(5), 407–417. doi: 10.1108/09513540710760183

Umble, E. J., Haft, R. R., & Umble, M. M. (2003). Enterprise resource planning: Implementation procedures and critical success factors. *European Journal of Operational Research, 146*(2), 241–257. doi: 10.1016/S0377-2217(02)00547-7

Varma, R. B. R., Umesh, I. M., Nagesh, Y. N., & Kumara Swamy, K. S. (2021). Digital transformation in higher education institutions – an overview. *International Journal of Applied Engineering Research, 16*(4), 278–282.

Verville, J., Bernadas, C., & Halingten, A. (2005). So you're thinking of buying an ERP? Ten critical factors for successful acquisitions. *Journal of Enterprise Information Management, 18*(6), 665–677. doi: 10.1108/17410390510628373

Vicedo, P., Gil, H., Oltra-Badenes, R., Merigó, J. M. (2020). Critical success factors on ERP implementations: A bibliometric analysis. In: Ferrer-Comalat J., Linares-Mustarós S., Merigó J., Kacprzyk J. (eds) *Modelling and Simulation in Management Sciences. MS-18 2018. Advances in Intelligent Systems and Computing*, vol 894. Springer, Cham. doi: 10.1007/978-3-030-15413-4_13.

Wanko, C. E. T., Kamdjoug, J. R. K., & Wamba, S. F. (2019). Study of a successful ERP implementation using an extended information systems success model in Cameroon universities: Case of CUCA. In: Rocha Á., Adeli H., Reis L., Costanzo S. (eds) *New Knowledge in Information Systems and Technologies. WorldCIST'19 2019. Advances in Intelligent Systems and Computing*, vol 930. Springer, Cham. doi: 10.1007/978-3-030-16181-1_68.

Weli, W. (2019). Enterprise resource planning implementation success factor (A case study in Atma Jaya Catholic University of Indonesia). *Journal of Theoretical and Applied Information Technology, 97*(11), 2988–3002.

Yokuş, G. (2022). Developing a guiding model of educational leadership in higher education during the COVID-19 pandemic: A grounded theory study. *Participatory Educational Research, 9*(1), 362–387. doi: 10.17275/per.22.20.9.1

3 Academic Training Contribution to the Development of a Leadership Profile

João Fernando Louro[1] *and Carolina Feliciana Machado*[1,2]

[1]School of Economics and Management, University of Minho, Braga, Portugal
[2]Interdisciplinary Centre of Social Sciences (CICS. NOVA.UMinho), University of Minho, Braga, Portugal

CONTENTS

3.1 INTRODUCTION

This chapter focuses on a critical review of the literature where the main theoretical currents around academic training in Portugal and its contribution to the development of a leadership profile are covered.

Human capital and leadership processes are variables with a very strong impact on organizational development, fundamentally due to the complex and unpredictable context that the labour market has been acquiring. The success and survival of an organization will depend on how its leaders lead it in face of the different challenges imposed by the context. This leadership action can influence people, negatively or positively, acquiring a crucial role in achieving the desired competitive advantage.

In this literature revision, entitled "The contribution of academic training to the development of a leadership profile", special focus will be given to themes related to training in higher education and its contribution to the development of leadership skills, of students who attend and attended different cycles of studies at the university level.

DOI: 10.1201/9781003021230-3

Throughout this critical literature review, it is proposed to actively reflect on the concepts of higher education in Portugal, the definition and development of the concept of competence and transversal competences, as well as the concept of leadership and its different approaches.

This reflection includes four sections, organized according to the following disposition. In the first section, a more generic allusion is made to academic training in Portugal and the development of skills, in the second the concept of competence, its definition and development is addressed, and in the third, the concept of transversal skills is presented, ending with the reference to the concept of leadership and its respective approaches.

3.2 HIGHER EDUCATION AND SKILLS DEVELOPMENT – THE CASE OF PORTUGAL

Higher education represents a vehicle for the creation and dissemination of scientific knowledge, in different areas of academic training, contributing on a large scale to the development of the economy. Through the evolution of labour markets, with the need to innovate and advances in terms of technology, we increasingly observe a growing need to incorporate highly specialized professionals with higher-level skills (Delors, 1996).

In Portuguese higher education, university and polytechnic education coexist, making this a binary system. The objective of university education focuses on research and knowledge creation, ensuring a solid foundation in terms of culture and science. In turn, polytechnic education is directed towards an applied investigation and practical development, aiming at preparing for the performance of professional activities. Both in university and polytechnic education, there are private institutions, in addition to public institutions.

Recently, and in line with the other member countries of the European Union (EU), we have witnessed the approximation of these two education systems. According to Lopes (2001), the emergence of the Bologna Process assertively defined a unique training model, regardless of whether the type of teaching is provided at a university or polytechnic institute. The Bologna Process led to the reorganization of courses into two main cycles, the first cycle – undergraduate, with a minimum duration of three years and oriented towards the labour market, and the second cycle – Master's, with a duration of two years and aimed at obtaining a specialization. In addition to these two cycles, there is also a third cycle – PhD, focused on research. Through the implementation of the Bologna Process, the first and second cycles were under the responsibility of university and polytechnic institutions, with the third cycle being exclusively the responsibility of universities. In addition to this restructuring and adoption of a model of organization of higher education in three cycles, the Bologna Process led to the organization of higher education courses based on the European system of transfer and accumulation of ECTS credits (European Credits Transfer System), defined in Decree-Law No. 42/2005, of 22 February, as well as assuming a clear need for transition from an education system based on the transmission of knowledge to a system based on the development of competences by the students themselves.

With these changes, the centrality of teaching processes passed from the teacher to the student, bringing with it, implications for teaching and assessment, demanding a different role for teachers (Clement, Gilis, Laga & Pauwels, 2008). The traditional training that was considered "as a function of the content and provision of the teacher, in a transmission logic" and which devalues "the person's role in the process of acquiring a new profile and takes him only as an 'instrument'" (Silva, 1999, p. 2), gives rise to a perspective of training based on the "proper logic of learning" (Altet, 1999, p. 13), in which the teacher is no longer a mere transmitter of knowledge and becomes a facilitator in the teaching-learning process (Zabalza, 2002). In this way, teaching is considered an intentional and interpersonal process that uses communication and the pedagogical context carried out "by the teacher as a means of provoking, favoring, achieving the learning of a knowledge or know-how" (Altet, 1999, p. 13).

This framework of the Bologna Process in higher education in Portugal is reflected in the amendment to the Basic Law of the Educational System (Law No. 49/2005, of 30 August) and in its statute regulated by Decree-Law No. 74/2006, of 24 March. Later on, some changes and adaptations are carried out with a view to improving and implementing the Bologna Process, through Decree-Law No. 107/2008 of 25 June.

In Portugal, access to higher education is subject to the numerus clausus system (each institution has a limited number of places). The institution may, if it so chooses, determine other selection criteria as well as a minimum score for those who wish to apply. It can also offer other entry alternatives through special access regimes, such as access for people over 23 years old to enter first cycle courses. For admission to second and third cycle courses, the institutions define for each course the criteria they intend to take to the competition, with the number of places being similarly limited.

The number of higher education graduates has grown year after year and has contributed to the increase in difficulties in the transition to the labour market, as well as the consequent concern with the skills acquired along the academic path (Bowers-Brown & Harvey, 2004).

According to Nasser and Abouchedid (2005), higher education institutions have begun to attach greater importance to the development of skills compatible with the demands of the labour market, especially with regard to transversal skills, those that are most valued by employers. These institutions have been carrying out initiatives aimed at developing the transversal skills of their students (Binks, 1996; Gammie, Gammie & Cargill, 2002), whether of an institutional nature (in specific institutions and on their own initiative), or through initiatives promoted by regional or governmental entities.

For authors such as Harvey, Moon, Geall and Bower (1997), the development of transversal skills should be an integral part of study programmes, as well as being part of the evaluation of the results of their students. Given that many of these skills are not exclusively developed in the curriculum, but through a set of informal learning processes, higher education institutions must also be able to recognize them as valid.

Despite the skills development initiatives that take place in higher education institutions show a trend towards training their graduates, with a view to improving employability rates, there are situations in which this intention allows the development of skills that enable the student to do them an asset throughout its formative process (Wittle & Eaton, 2001). In addition to these skills, allowing the student to achieve greater academic success, they will become an asset in the different professional contexts they integrate after the end of their study cycle.

Higher education training has a great impact on what are the trajectories of training and professional activity of individuals. The dynamics and structural particularities of institutions can lead to the development of actions and conceptions, favourable or limiting of autonomy, adaptation or transformation, according to the guidelines that each higher education institution circumscribes (Dias Sobrinho, 2010). The educational process of higher education institutions should not only be considered from the point of view of the teaching and learning process, but also in their social, economic, ethical, cultural and political dimensions. Thus, higher education should focus on training individuals beyond their technical skills, emphasizing the training and development of a profile committed to ethical, political, economic and sociocultural issues (Dias Sobrinho, 2010; Marinho-Araujo, 2004; Vieira & Marques, 2014).

According to Rué (2007), it is imperative that higher education institutions invest in the development of the curriculum and training actions that allow the development of skills in higher education. However, this implies that training adjusts, resorting to new models of learning, teaching and evaluating (Rué, 2007). Regarding the importance of transversal skills in higher education, a group of researchers also state that higher education institutions should evaluate and analyze "the review of the pedagogical methods and techniques used, as well as the study plan of some courses, with more subjects to choose from, if the concern is guided by training in transversal skills" (Cardoso, Estêvão & Silva, 2006, p. 181). It is up to higher education institutions to equate mechanisms capable of providing their students with experiences outside the University, such as through internships during the course and not just at the end of the course (Cardoso et al., 2006).

The need to transfer what has been learned to different contexts is, nowadays, one of the great challenges faced by professional training practices. In higher education training, there are few situations in which it is taught to reflect, critically and autonomously, in the face of situations presented and the different and possible modalities of action to be developed. In this sense, it is relevant to consider with special emphasis the transversal competences and their respective inclusion in the processes of higher education, in a more evident and strong way in their curricula. In order to make the development of competences possible in the training process associated with higher education, it is essential to have methodological references supported by the categorization of competences. Thus, the reorganization and development of curricular and pedagogical plans by higher education institutions is essential, not only in the sense of responding assertively to the assumptions outlined by the Bologna Process, responding to the requirements emanating from the EU, but also in the sense of to make it possible to improve the quality of academic training and the empowerment of its students.

3.3   COMPETENCE: CONCEPT DEFINITION AND DEVELOPMENT

The concept of competence was first explored during the Taylorist movement, but it was in the early 1980s that it began to gain strength in the world of work through its overlapping with the concept of qualification (Parente, 2004; Stroobants, 2006).

At this point, the ability to effectively mobilize knowledge and skills for a more efficient work performance, whether personal or acquired, begins to be valued, to the detriment of school certification.

With the multiplicity of tasks and the huge accumulation of functions, technological evolution and the changing nature of the labour market, the human resources of organizations began to aggregate and develop a whole set of skills that could be put into practice in other contexts of work, enabling greater versatility in the performance of different functions and the possibility of embracing other professional paths (Guichard & Huteau, 2002; Michael, Arthur & Judith, 1999).

Faced with this context of uncertainty and permanent transformations, according to Boterf (2003), organizations look for individuals capable of mobilizing knowledge between different professional contexts. This search, despite not presenting itself as a direct solution to the possible concern with the risks of unemployment, reveals that "a set of validated skills and a proven ability to enter into learning processes will have an appreciable advantage in the labor market" (Boterf, 2005, p. 16), making the individual capable of defending their employability in an autonomous and proactive way.

Etymologically, the word competence has its origin in the Latin, *competens*, and means "what goes with, what is adapted to" as stated by Boterf (2003, p. 53). For the author, competence brings the ability to analyse to the ability to solve problems, in a work context that regularly needs adaptation. The concept of competence has a polysemic character, which reveals some lack of consensus as to its meaning (Hoffmann, 1999; Stroobants, 2006). Its wide use in different areas of knowledge, such as psychology, politics, as well as management and education, translates into one of the main reasons for the lack of consensus in its definition (Boterf, 1997).

Given the polysemy of its meaning, Stroobants (2006) refers to competence as the ability to mobilize different knowledge, knowing how to do and knowing how to be, to solve problems over time and in different contexts. Based on a critical perception of its potential and capabilities, and aware of its limitations and difficulties, Gonçalves (2006) refers to the concept of competence as the ability to achieve answers, results and solutions, in different dimensions and contexts. Also, in the same line of thought "being competent is increasingly being able to manage complex and unstable situations" (Boterf, 2005, p. 18).

Cardoso et al. (2006), in their publication, tell us that the complexity of this concept reflects the effect of the orientations of the capitalist economic system that imposes, on organizations and their human resources, a greater capacity for result orientation, with the clear objective of responding to growing globalization and the evolution of the labour market, based on strong competitiveness traits. Being competent requires knowing how to act and react in an adjusted way to the sudden and unexpected event. Implies the ability to "self-regulate its actions, knowing how to count not only on its resources, but looking for complementary ones, being in a position to transfer them and (re)invest its skills in a different context" (Castro, 2007, p. 7).

Competence as a "validated – operational – know-how" is described by Meignant (2003, p. 281/282) as the ability to carry out a singular intervention in the face of an unprecedented and unknown problem, which "allows to immediately mobilize theoretical, procedural, knowledge experimental, empirical, social, cognitive, to find an innovative answer to a situation that could not be entirely foreseen by the study offices. This combinatorial ability is at the heart of competence".

According to Rogiers and De Ketele (2004), competence is understood as a concept capable of mobilizing knowledge between different contexts that allows "spontaneously – to apprehend a situation and respond to it more or less pertinently" (p. 45), stating that all the competence that a professional possesses should be taken into account in a perspective of integrating knowledge that was developed in learning processes in the most varied contexts (formal, non-formal, informal).

For Stroobants (2006), in short, competence is understood as the ability to mobilize knowledge, knowing how to do and knowing how to be, in different contexts. Competence is not limited to what is external and observable and has a certain particularity, since its tacit nature leads us to a non-formal and informal dimension of competences, in which, implicitly, as a result of actions external to the traditional education system, is reflected in practices in the work context.

Regarding the evolution of this concept within education and training practices, it involves new requirements and considerations. Great importance is given to the practical aspect (knowing how to do) in conjunction with theoretical and technical contents, in addition to its obvious manifestation in relation to pedagogical practices (Parente, 2004). Even so, the acquisition and development of skills is not solely and exclusively the result of the formal dimension of the teaching and learning system, as mentioned earlier. In their social dimension, family, work and all practices and experiences should be considered widely rich contexts for the acquisition and development of skills, despite their informal nature (Boterf, 2003; Pires, 2002). Pires (2002), in his systemic approach, defends that competences result from the articulation between knowledge of different nature, mobilized in a certain action, through a process that is integrative, contextualized and objective. We can thus say that competences are the result of complex combinations of attributes such as knowledge, behaviour, values, attitudes and cognitive strategies, mobilized in order to answer to complex situations that individuals experience in different contexts. This process of skills development assigns shared responsibilities to the articulation of formally acquired learning and the individuals' life path (non-formal and informal).

From a constructivist perspective, the development of competences is always a process contextualized in social and personal terms, which develops progressively, in which context and motivation acquire crucial importance (Trepós, 1996). During this evolution, Wittorsky (1998) tells us that competences are developed through a mechanism that calls for a constantly critical and reflective exercise, during the process that exists between the activity and its conceptualization.

Thus, it is entirely pertinent to consider the process of acquiring and developing skills as the result of a cluster of experiences, training and work, and not as an activity exclusively parallel to or prior to work. We can say that competences are not innate qualities to the subject, they are not taught or transmitted, but rather the result of a set of experiences that the individual has experienced, acquired and mobilized.

3.4 TRANSVERSAL SKILLS

The growing importance of transversal skills is directly linked to the particularities to which the work context has been subjected, as a result of the great unpredictability and changeability of the labour market. Given its impact on human resource management, transversal skills have increasingly acquired greater centrality in management processes, especially in what we call competency-based management. Investment in skills development processes should "consider the great centrality of transversal skills (organization, animation, development skills) and of know-how (sociability, adaptability, charisma, sense of communication)" (Estêvão, 2012, p. 117).

As might be expected, if the concept of competence is not unanimous and presents some concerns in terms of its conceptualization, the concept of transversal skills confronts us with the same complexities. Tien, Ven and Chou (2003) list in their work a set of names that this concept presents, according to different countries and organizations, namely:

- "Employability skills" – United States of America
- "Core skills" – United Nations
- "Key competencies" – Australia
- "Core skills/key skills" – Great Britain
- "Employability skills" – Canada
- "Basic competencies" – Taiwan

In Portugal, there are several designations used to refer to this concept, such as (Cardoso et al., 2006, p. 36):

- "Essential skills"
- "Generic skills"
- "Key skills"
- "Nuclear competences"
- "Transferable skills"

Currently, a widely used term is that of soft skills. By soft skills, we can consider the set of behavioural, interpersonal and transversal skills to different professional areas that are fundamentally characterized by non-technical skills (Klaus, 2007) that help the persons to improve their professional performance, such as communication, teamwork, negotiation skills, leadership, among others (Seth & Seth, 2013). This knowledge encompasses a "set of skills that, as the name indicates, are transversal to different professions/professional activities and that facilitate the employability (in the broad sense here) of those who possess them" (Cardoso et al., 2006, p. 37).

Soft skills can be understood as transferable skills that contribute to competent performance in different areas (Gibbons-Wood & Lange, 2000), contributing to good "performance in all sectors and at all levels" (Chadha, 2006, p. 19). These can be understood as capacities, attitudes and skills of the individual that lead the person to respond effectively in different professional situations, being transferable between contexts throughout life, as defended by Moreno (2006). As described by the International Labor Organization (OIT, 2002), transversal skills are those that appear as basic and common to different work activities, enabling the correspondence between different professional profiles, or from a set of curricular modules to others.

For authors, such as Jardim and Pereira (2006), transversal skills include three dimensions: intrapersonal, interpersonal and professional. Thus, individuals with these three well-developed dimensions will be able to manage their personal attributes, which sustain the relationship with the other and influence their professional performance; relate to other individuals; perform their professional functions, through the practical application of specific knowledge of their field of work.

According to Tien et al. (2003) and Kearns (2001), there are two major approaches around transversal skills. The first, called the broad approach, more suited to the North American context, considers the existence of a more holistic set of skills, which includes skills related to employability and lifelong learning, in addition to basic skills, ethics and value judgment and personal attributes. Regarding the second, the restricted approach, characteristic of countries like Australia and the United Kingdom, its consideration is carried out from a more instrumental perspective, emphasizing these same skills in the work context. This second approach is a model more influenced by the competence-based training approach (competency-based training).

Mansfield (2003) tells us that transversal skills arise, in an organizational context, as a reaction to the occupational model of competences, based on functional analysis. The author argues that, currently, the organizational context requires a broad set of skills, leading to the need for an articulation and balance between the specific skills associated with a given professional activity and the transversal skills, in order to answer to the different requests from the work context.

Despite the existence of several definitions around soft skills, we can consider two key elements. These can be the result of an activity or discipline but can be used in different domains (Ceitil, 2007) and can be applied in different contexts and situations (Allen, Ramaekers & Van der Velden, 2005; Ceitil, 2007).

In the literature, there are several authors who point to their contribution to professional success, in the sense that they make it possible for individuals to acquire and develop the necessary tools to face the demand, competitiveness and uncertainty of the labour market (Cardoso et al., 2006; Rey, 2002; Rocha, Gonçalves & Vieira, 2012; Van der Klink, Boon & Schlusmans 2007).

3.5   THE CONCEPT OF LEADERSHIP AND ITS APPROACHES

The ability to lead has always had a great impact throughout the evolution of humanity, being a subject of interest for thousands of years. However, scientific research in this area only began in the 20th century (Fry, 2003) and since then, as Bass (1990) refers, the definitions of leadership are almost as many as the number of people who have dedicated themselves to its study.

Thus, leadership has aroused the interest of a large number of researchers (Cunha, Rego, Cunha & Cabral-Cardoso, 2007). In the current context, specifically within organizations, the centrality of this concept is extremely important, since, according to Syroit (1996), given the incomplete nature of organizational design, in terms of regulating the behaviour of its individuals, leadership acquires special importance in the balance, control and stability of a given organization.

Regarding its definition, according to Neves (2001), and despite the large number of literature produced around the concept of leader and leadership, the results of the research are not always an asset for the definition of the concept, ending up, many times, relating to the concepts of power, authority and management. According to Yukl (2013), and in the same line of thought, based on the concept of leadership, there is a certain association with the concepts of power, authority and influence. Even with the large amount of these studies and their contribution to clarifying this concept, there is no consensual definition about it.

Some authors argue that, if it were possible to reach a unanimous definition in relation to the concept of leadership, it would be better understood and operationalized (Kort, 2008), because, and according to Oliveira (2008, p. 38), "realizing the role and function of leadership seems to be the most important intellectual task of this generation and leading the most needed competence". For Kort (2008), the existing definitions differ in the way leaders view the concept, the way they motivate their followers and how they lead to the interests of the group or organization.

For Daft (2010), leadership occurs between people, using influence and with a view to achieving a certain objective. According to Syroit (1996), all definitions of leadership imply that, within a given group, one or more members can be identified as leaders, and these members are distinguished by having a specific set of characteristics and the others present themselves as "followers" or "subordinates", creating a hierarchical structure within this group. The aforementioned author states that, in certain definitions, leadership is seen as a process of interaction between different members and that leaders are responsible for influencing their followers, through certain orientation towards results and/or objectives.

Leadership is a concept that can be understood as the "ability of an individual to influence, motivate or enable others to contribute to the organization's effectiveness and success", as defended by Wilderom et al. (1999, p. 184). Considering leadership as an element of a management process, we can say that the complexity of this process is only effective when its leadership is effective. For authors such as Chiavenato (2001), Azevedo and Costa (2004) and Uhl-Bien and Arene (2018), leadership is the ability to lead people to achieve an objective, project and/ or action, influencing them to produce and be motivated to achieve the organizational mission.

The large number of studies on leadership issues consequently reflects the huge variety of theories and models. According to Syroit (1996), this abundance of models and theories makes the attempt to explain and define the concept of leadership more complicated.

Given this multiplicity, three approaches are accepted around the concept of leadership, developed over the years, namely (Cunha et al., 2007; Yukl, 2013):

1. *Traits approach:* Characterize and identify the individual attributes of those who exercise leadership roles, in order to establish a distinction between leaders and non-leaders;

2. *Behavioural approach*: Centred on the analysis of what leaders do in carrying out their functions, seeking to determine the behaviours that are most related to organizational effectiveness;

3. *Situational/contingent approach*: Privileges the observation of environmental factors that condition the results of leadership.

However, in recent years, new approaches and perspectives have emerged that relate the leadership process to the charisma and transformational capacity of leaders in their relationship with employees (Santos & Caetano, 2007). In this way, we observe a concern with relating behavioural aspects, personality traits and the type of interaction established between individuals in a given organization. Thus, we will also approach this fourth set of theories called integrative.

Table 3.1 summarizes the four main trends in different approaches to leadership, presenting their assumptions and associated theories.

As we can see, the concept of leadership has been broadly addressed by several authors, and along this conceptual evolution, several theories, approaches and styles of leadership emerged with the aim of explaining this phenomenon, despite the lack of a universal and globally accepted definition. As mentioned by Burgoyne,

TABLE 3.1
Leadership Approaches

Approaches	Assumption	Associated Theories
Traits	Leadership is an innate attribute	• Personality Traits Theory
Behavioural	Leadership effectiveness is related to the type of behaviour a leader adopts	• Ohio Studies • Michigan Studies • Likert Rating • Blake and Mouton • Management Grid
Situational or contingent	Leadership effectiveness is influenced by different contexts	• Continuum Leadership Theory • Power-Influence Theory • Fiedler's Contingency Theory • Situational Theory by Hersey and Blanchard • Objective-Path Theory • Normative Decision-Making Theory • Leadership Surrogate Theory • Multiple Links Theory • Cognitive Resource Theory
Integrative	Leadership will depend on the leader's vision	• Charismatic Leadership • Transactional Leadership • Transformational Leadership

Source: Adapted from Cunha et al. (2007).

Boydel & Pedler (2004), after so many decades of research, the concept of leadership remains undefined.

3.6 CONCLUSIONS

Academic training presents itself as a preponderant stage in the development of skills and abilities, both technical and specific and transversal. The training and development of individuals is one of the key pieces in the strategic management of organizations, contributing not only to achieving competitive advantage within the organization, but also to the personal and social development of individuals.

With this brief literature review, we intend to reflect on the main theoretical currents of academic training in Portugal and its contribution to the development of leadership skills, of those who attend or attended the university in different cycles of studies.

From the literature review carried out, we can conclude that, in general, higher education, in addition to representing a vehicle for the creation and dissemination of scientific knowledge, has gradually been giving special emphasis to the development of skills compatible with the demands of the labour market, namely transversal skills. At the same time, the labour market, due to its changing and fickle nature, requires individuals to be able to mobilize a wide range of skills efficiently and effectively, between different work contexts, contributing to a greater need for development of those that are the transversal skills and the consequent enhancement of them. The role of the leader within the organizational environment acquires special importance, and this should be the main driver of ambition, understanding, confidence, positive attitude and efficient management of its followers and employees. It must also have the ability to inspire and create new leaders through innovation, creativity and autonomy, without neglecting the organization's collective vision, mission and values. The concept of leadership has been addressed by several authors, and along its conceptual evolution, several theories, approaches and styles of leadership emerged.

Although this brief conclusive synthesis results only from the treatment of theoretical concepts in the literature, it still allows us to draw relevant fundamental conclusions for a better understanding of the role of higher education and, consequently, of universities, in the promotion and development of transversal skills and leadership in its students.

REFERENCES

Allen, J., Ramaekers, G., & Van der Velden, R. (2005). Measuring Competencies of Higher Education Graduates. *New Directions for Institutional Research*, 126, 49–59.

Altet, M. (1999). *As Pedagogias da Aprendizagem*. Lisboa: Instituto Piaget.

Azevedo, I. & Costa, S. (2004). *Secretária – um guia prático*. (4ª ed.). São Paulo: Senac.

Bass, B. (1990). *Handbook of Leadership: A Survey of Theory and Research*. New York, NY: Free Press.

Binks, M. (1996). Enterprise in Higher Education and the Graduate Labour Market. *Education and Training*, 38(2), 26–29.

Boterf, G. (1997). *De la compétence à la navigation professionnelle*. Paris: Les Éditions d''Organisation.

Boterf, G. (2003). *Desenvolvendo a competência dos profissionais*. São Paulo. Artmed.

Boterf, G. (2005). *Construir competências individuais e colectivas. Resposta a 80 questões.* Lisboa: Edições Asa.

Bowers-Brown, T. & Harvey, L. (2004). Are There Too Many Graduates in the UK? A Literature Review and an Analysis of Graduate Employability. *Industry & Higher Education*, 18(4), 243–254.

Burgoyne, J., Boydel, T., & Pedler, M. (2004). Suggested Leadership Development. *People Management*, 10(4), 46.

Cardoso, C. Estevão, C., & Silva, P. (2006). *Competências transversais dos diplomados do ensino superior: Perspetiva dos empregadores e dos diplomados.* Guimarães: Tecminho/Gabinete de formação continua.

Castro, J. (2007). Dupla certificação. *Formar, Revista dos Formadores*, 60, 4–11.

Ceitil, M. (2007). *Gestão e Desenvolvimento de Competências.* Lisboa: Sílabo.

Chadha, D. (2006). A Curriculum Model for Transferable Skills Development. *Engineering Education*, 1(1), 19–24.

Chiavenato, I. (2001). *Teoria Geral da Administração.* (6ª ed.). Rio de Janeiro: Elsevier. p. 19.

Clement, M., Gilis, A., Laga, L., & Pauwels, P. (2008). Establishing a Competence Profile for the Role of Student-Centred Teachers in Higher Education in Belgium. *Research in Higher Education*, 49(6), 531–554.

Cunha, M., Rego, A., Cunha, R., & Cabral-Cardoso, C. (2007). *Manual de Comportamento Organizacional e Gestão.* Lisboa: Editora RH.

Daft, R. (2010). *Management.* (9ª ed.). Mason, OH: South-Western Cengage Learning.

Decreto-Lei n° 42/2005, de 22 de fevereiro, 2005.

Decreto-Lei n° 74/2006, de 24 de março, 2006.

Decreto-Lei n° 107/2008, de 25 de junho, 2008.

Delors, J. (1996). *Educação um tesouro a descobrir.* (2ª ed.). Lisboa: Edições Asa.

Dias Sobrinho, J. (2010). Democratização, qualidade e crise da educação superior: Faces da exclusão e limites da inclusão. *Educação e Sociedade*, 31(113), 1223–1245.

Estêvão, C.V. (2012). *Políticas & valores em educação. Repensar a educação e a escola pública como um direito.* Braga: Universidade do Minho, Instituto de Educação: Edições Húmus.

Fry, L. (2003). Toward a Theory of Spiritual Leadership. *Leadership Quarterly*, 14, 693–727.

Gammie, B., Gammie, E., & Cargill, E. (2002). Personal Skills Development in the Accounting Curriculum. *Accounting Education*, 11(1), 63–78.

Gibbons-Wood, D. & Lange, T. (2000). Developing Core Skills: Lessons from Germany and Sweden. *Education and Training*, 42, 24–32.

Gonçalves, C. (2006). *A família e a construção de projectos vocacionais de adolescentes e jovens.* Dissertação de Doutoramento. Porto: Faculdade de Psicologia e Ciências da Educação da Universidade do Porto.

Guichard, J. & Huteau, M. (2002). *Psicologia da orientação.* Lisboa. Instituto Piaget.

Harvey, L., Moon, S., Geall, V., & Bower, R. (1997). *Graduate's Work: Organisation Change and Students' Attributes.* Birmingham: Centre for Research into Quality Together with the Association of Graduate Recruiters.

Hoffmann, T. (1999). The Meanings of Competency. *Journal of European Industrial Training*, 23, 275–286.

Jardim, J. & Pereira, A. (2006). *Competências pessoais e sociais: Guia prático para a mudança positiva.* Porto: Edições ASA.

Kearns, P. (2001). *Generic Skills for the New Economy.* Austrália: National Center for Vocational Education Research (NCVER).

Klaus, P. (2007). *The Hard Truth about Soft Skills.* New York, NY: HarperCollins.

Kort, E. (2008). What, after All, Is Leadership? "Leadership" and Plural Action. *Leadership Quarterly*, 19, 409–425.

Lei n° 49/2005, de 30 de agosto, 2005.

Lopes, R. (2001). Actuais tendencias do ensino superior politecnico na Europa. *Politecnica*, 2(2), 10.

Mansfield, B. (2003). Competence in Transition. *Journal of European Industrial Training*, 8(2/3/4), 296–309.

Marinho-Araujo, C. (2004). O Desenvolvimento de Competências no ENADE: A mediação da avaliação nos processos de desenvolvimento psicológico e profissional. *Avaliação – Revista da Rede de Avaliação Institucional da Educação Superior – RAIES, Unicamp, Campinas*, 9(4), 77–97.

Meignant, A. (2003). *A gestão da formação*. (2ª ed.). Lisboa: Publicações Dom Quixote.

Michael, B., Arthur, I., & Judith, K. (1999). *The New Careers, Individual Action and Economics Change*. London: Sage.

Moreno, M. (2006). *Evaluación, balance y formación de competencias laborales transversales: propuestas para mejorar la calidad en la formación profesional y en el mundo del trabajo*. Barcelona: Laertes Educación.

Nasser, R. & Abouchedid (2005). Graduates's Perception of University Training in Light of Occupational Attainment and University Type. *Education and Training*, 47(2), 124–133.

Neves, J. (2001). O processo da liderança. In Ferreira, J., Neves, J., & Caetano, A. (Eds.), *Manual de Psicossociologia das Organizações*. Lisboa: Editora McGraw-Hill.

Oliveira, A. (2008). *Liderança e espiritualidade nas organizações: um estudo exploratório*. Braga: Universidade do Minho.

Organização Internacional do Trabalho (OIT). (2002). *Glossário de termos técnicos – Certificação e avaliação de competências*. Brasília: OIT. Available at: http://www.ilo.org/brasilia/publicacoes/WCMS_221528/lang–pt/index.htm, accessed in 20 December 2021.

Parente, C. (2004). Para uma análise de gestão de competências profissionais. *Sociologia: Revista da Faculdade de Letras da Universidade do Porto*, 14, 299–343.

Pires, A. (2002). *Educação e Formação ao Longo da Vida: análise crítica dos sistemas e dispositivos de reconhecimento e validação de aprendizagens e de competências*. Tese de Doutoramento. Lisboa: FCT/UNL.

Rey, B. (2002). *As competências transversais em questão*. São Paulo: Artmed Editora.

Rocha, J., Gonçalves, C., & Vieira, D. (2012). Competências transversais: perceção de estudantes do 1º ano do ensino superior. *Apoio psicológico no ensino superior: um olhar sobre o futuro: Actas do II congresso RESAPES-AP*, 196–206. ISBN 978-989-97851-0-6.

Rogiers, X. & De Ketele, J. (2004). *Uma pedagogia de integração: Competências e aquisições no ensino*. Porto Alegre: Artemed.

Rué, J. (2007). *Enseñar en la Universidad: El EEES como Reto para la Educación Superior*. Madrid: Narcea.

Santos, J. & Caetano, A. (2007). Liderança Transformacional: Predição de eficácia dos líderes e da satisfação percecionada pelos subordinados. *Percursos da Investigação em Psicologia Social e Organizacional*, 3, 179–192.

Seth, D. & Seth, M. (2013). Do Soft Skills Matter? – Implications for Educators Based on Recruiters' Perspective. *The IUP Journal of Soft Skills*, VII(1), 7–20.

Silva, E. (1999). *Olhares sobre a Formação: Concepções, Práticas e Paradoxos*. Braga: Universidade do Minho, Instituto de Educação e Psicologia (Texto policopiado).

Stroobants, M. (2006). Competência. *Laboreal*, 2(2), 78–79.

Syroit, J. (1996). Liderança organizacional. In Marques, C. & Cunha, M.P. (Eds.), *Comportamento Organizacional e Gestão de Empresas* (pp. 237–277). Lisboa: Publicações D. Quixote.

Tien, C., Ven, J., & Chou, S. (2003). Using Problem-Based Learning to Enhance Student's Key Competencies. *Journal of American Academy of Business*, 2(2), 454–459.

Trepós, J. (1996). *Sociologie de la compétence professionnelle*. Nancy: Presses Universitaires de Nancy.

Uhl-Bien, M. & Arene, M. (2018). Leadership for Organizational Adaptability: A Theoretical Synthesis and Integrative Framework. *The Leadership Quarterly*, 29, 89–104.

Van der Klink, M., Boon, J., & Schlusmans, K. (2007). Competências e ensino superior profissional: presente e futuro. *Revista Europeia de Formação Profissional*, 40, 72–89.

Vieira, D. & Marques, A. (2014). *Preparados para trabalhar? Um estudo com diplomados do Ensino superior e empregadores*. Lisboa: Forum Estudante.

Wilderom, C.P.M., House, R., Hanges, P., Ruiz-Quintanilla, S., Dorfman, P., Javidan, M., & Dickson, M. (1999). Cultural Influences on Leadership and Organizations: Project 20 GLOBE. In Mobley, W.H., Gessner, M.J., & Arnold, V. (Eds.), *Advances in Global Leadership* (pp. 171–233). Stamford, CN: JAI Press.

Wittle, S. & Eaton, D. (2001). Attitudes towards Transferable Skills in Medical Undergraduates. *Medical Education*, 35, 148–153.

Wittorsky, R. (1998). De la fabrication des compétences. *Education Permanente*, 135, 57–69.

Yukl, G. (2013). *Leadership in Organizations* (8ª ed.). Boston, MA: Pearson.

Zabalza, M. (2002). *La Enseñanza Universitária: El Escenario e sus Protagonistas*. Madrid: Narcea.

4 Higher Education Policy and Widening Participation

An Overview

Vlasios Sarantinos
Faculty of Business and Law, University of the
West of England, Bristol, United Kindgom

CONTENTS

4.1   INTRODUCTION

One of the main concerns of education policy in the UK higher education (HE) sector has historically been the issue of widening participation (WP) – enabling access to University from social groups that are underrepresented, mainly due to socioeconomic reasons e.g. people from Black and Asian minorities (McCaig, 2018). This chapter will seek analyse how WP policy emerged in the past few decades; what have been the main developments and influences that have shaped both policy formulation and enactment. Building on the findings, the effort will be to try and crystallise a clear understanding of the success, challenges and present status of the field.

4.2   WIDENING PARTICIPATION POLICY: A HISTORICAL ROUTE

The history of accessing HE in the UK has a long-standing past. Going back to the 1950s and 1960s, the perception of the role universities had to play and their importance in society begun to transform, with voices suggesting that institutions of HE should not be an exclusive privilege to the elite, but instead more easily accessible to the wider public, particularly given the capacity and potential to contribute to the nation's development (Stevens, 2004). Perhaps, one of the most influential thought

DOI: 10.1201/9781003021230-4

pieces of those early years has been the Robins report, produced by the Committee for HE which laid the foundations for the driving forces behind HE's transformation as a landscape (Burke, 2012). For the first time, there was a clearly outlined call for the relevant stakeholders to consider the entire scope of University education, with particular emphasis on 'courses of higher education should be available for all those who are qualified by ability and attainment to pursue them and who wish to do so' (CHE, 1963:8). Without a doubt, this principle resonates in terms of rhetoric to this very day and can be registered in even the more contemporary endeavours to increase access to HE.

To a large extent, improving opportunities for the populace to continue with their studies was geared towards improving social mobility and cohesion (Macleod, 1996), but also to contribute to skills development, employment and overall growth of the economy (HEFCE, 2001). Indeed, both appear even to-date to be very influential in shaping the motivation behind WP. The former has also evolved to encompass the need to address social injustices and improve equality, alongside providing a plat-form for progression and growth for groups of underprivileged and underrepresented groups of the society (Evans et al., 2019). From an education policy perspective, although this has been very clear documented in relevant documents e.g. (BIS, 2014; DfES, 2003), this could potentially signify larger contextual links, aligning with the wider policy decision-making to engineer push for equality and cater for the needs of the labour market (Jones & Thomas, 2005). This is also evident in the framing, interpretation and implementation of the policy within the sphere of education policy across the different levels, an aspect demonstrates the inherent complexity across all stages of policy construction and manifestation (Trowler, 2002).

As part of championing greater inclusivity and access to HE, the number of uni-versities increased dramatically over the past decades in order to ensure there were sufficient positions for students to be absorbed and as also to respond to the need for increased skilled labour (Machin & Vignoles, 2006). The link between universities and the labour market is an aspect that would suggest a connotation with the broader contextual factors, for example the need to address skills shortages. Despite the con-tinued expansion of the sector, although some improvements surfaced, particularly in the form of more students from underprivileged background having a greater pos-sibility to attend university than in the past, proportionately the rise has been greater for more affluent groups (Galindo-Rueda et al., 2004). Similarly, people from Asian or Black Ethnic groups are less likely to enrol onto a University course than their White counterparts (Boliver, 2013). Another consequence of the number of institu-tions and the increased student numbers was the unavoidable change in the funding system which led to the introduction, and gradual increase of tuition fees that have come to burden the students (Callender, 2013; Galindo-Rueda et al., 2004).

4.3   WHAT DRIVES WP POLICY?

In order to unravel this nuances of how the education policy around WP evolved, it would be useful to look more carefully at the root causes that induced the very drives behind it. As a starting point, it would be useful to examine the nature of the issue and how this has been depicted in the various relevant texts. As mentioned earlier, from the days of the Robin's report, there would appear to be an enhanced emphasis on remedying injustices and inequalities within the society and supporting the groups that have been historically marginalized. Equally, economic and financial reasons are also evident, through the focus of upskilling the workforce and remaining competitive at the global stage (Machin & Vignoles, 2006). Perhaps another interesting notion, again captured within the Robin's principle is the need for meritocracy, with emphasis on ability and attainment.

HEFCE (2001) posits that it is quite difficult to frame precisely what a set definition of WP might be, and adheres to the generic term of facilitating access to under-represented and underprivileged groups, but with the added dimension of ensuring that they will be able to transit into academia successfully. It also suggests that there appears to be significant variance amongst institutions in relation to how this strategy is pursued and applied. In similar vein, Hubble and Bolton (2018) concur with the broad criteria of who is targeted by the drive to increase access to HE, as people belonging in lower-income household groups and from other under-represented parts of the society, with emphasis on people from Black and Asian Ethnic minorities. The authors also suggest that part of the policy is also the help progress within the sector, enhance students' graduate outcomes and strengthen employability. In terms of how the policy is enacted and monitored, that is done in three levels, directly through institutions, through a public regulator (office for fair access) and through the funding mechanisms controlled by the HEFCE. Martin (2018) notes although progress has been made over the years in terms of reducing the gap, with participation in the 1980s being between 10 and 15% to almost triple now, standing at 45%, a more careful look in statists reveals a less satisfactory picture as there is tremendous difference between regional and local areas in terms of participation, suggesting that the factors that create the barriers in the first place might well still hold strong.

A study commissioned by JSCM for HEFCE & UUK in 2004, indicates that despite the emphasis and attention to the phenomenon over the preceding decades, there is still very little substance behind a clear definition of what is means and entails. This is in itself problematic, as the loose use of umbrella terms such as under-represented and underprivileged can be subjective, vary from area-to-area and can lend to creating a complex landscape. Clarke (2018) drawing on empirical data from Higher Education Statistics Agency argues even though students who are targeted to enter university on account of being disadvantaged might be able to do so, once they commence their studies, they face significant issues and challenges, leading to high levels of discontinuation, substantially higher than non-disadvantaged students. On the issue of employability, the author also notes that even if they do manage do successfully complete their studies and graduate, working-class new entrants in the labour market are paid less than their more affluent counterparts, with the former group also facing more obstacles in terms of promotion and progression.

Although the previous discussion is by no means exhaustive, a number of interesting elements emerge as to how the policy is portrayed. Firstly, it still remains at the forefront of all actors, from national to local level without having abated in terms of drive and attention. While some progress seems to have been made, there seems to be a consensus that there is need for significantly more progress and that potentially the wider root causes that create the very inequalities are not addressed sufficiently.

Within the field, there also seems to be a rather intricate arena with multiple stake-holders and confluence. Correspondingly, the next parts will try to analyse the extent to which WP has been pursued within the UK context, particularly focusing on what steps have been undertaken to operationalize the policy, how these were carried out and with what impact. The analysis will also try to highlight the underlying political contours and also consider the interplay between different actors.

In the late of 1980s, an increase to the number of school leavers with high enough results to carry onto HE, created an additional impetus to revisit the number of institutions with degree-awarding powers (Carpentier, 2018). Consequently, these demographic trends provided additional bolstering to the call to augment the num-ber of providers bearing University status, in part with the priorities of continuing with the WP agenda (Machin & Vignoles, 2006). In a similar fashion, more recently the government decided to lift the cap on student numbers for the 2014–2015 aca-demic year, enabling therefore universities to recruit without a limit student with home status (Riddell, 2015). Indeed, as Jo Johnson, the Education Minister of that period himself stated 'We're nearing the end of the university application cycle and the latest data tells of a bumper year. Record numbers of applicants accepted onto their first choice of course. Record acceptances for young people from disadvantaged backgrounds. All this is possible because, for the first time, we have lifted the cap on student numbers. More than 50 years after the Robbins report made the case for university expansion, we are the first Government to live up to its guiding prin-ciple' (Johnson, 2015). We can therefore tease out a few observations in terms of the mechanisms government in the UK have adopted in terms of realizing the WP agenda. Firstly, the most typical response appears to be an expansion of the sector in size, either in terms of the number of institutions or in relation to possible student population. Secondly, it is quite telling that the very same principles that guided the Robbins committee in the 1960s still hold such sway in the more than 50 years later. Nevertheless, the natural expansion of the sector both in terms of size and scope required an equal increase in terms of investment to maintain the necessary stan-dards and quality needed, which given that the funding provided by the government remained the same, led to the inevitable introduction and subsequent increase of the tuition fees levelled on the students (Machin & Vignoles, 2006; Murphy et al., 2017).

4.4   TUITIONS FEES AND WIDENING PARTICIPATION: A QUESTION OF POLITICS

Up until 1998, the students attending HE paid no fees, as the system was wholly sub-sided by the government through its relevant agencies. Nevertheless, the New Labour party who was at the government then, decided to introduce a 1,000 pounds fee per annum which was meant to be means-tested (Pennell & West, 2005). Murphy et al. (2017) examining the reasons that led the Labour government to press for the intro-duction of fees in the late 1990s, postulate a combination of increasing demand for HE, large number of institutions but also call for more equitable and fair allocation of the funding, given that all students invariably of their financially background ben-efited the same which some viewed as favouring unjustly students from more affluent backgrounds. Correspondingly, part of the argument for introducing fees was also to ensure resources would be channelled to support less privileged students and mitigate the barrier financial constraints might impose. The fees further improved in a two-stage process, trebling in 2004 to 3,000 pounds again by the incumbent Labour government and then trebling again in 2012 at 9,250 by the coalition government (Murphy et al., 2017).

Although the possibility to redeploy the resources available more strategically in order to support access and attainment of students in need might appeal as rhetoric, the evidence is less encouraging. Although some authors do suggest that there has been some improvement, and indeed students have benefitted, the introduction of fees was perhaps more the aftermath of rational economic thought, in order to address the challenges of the sector (Brown & Carasso, 2013; Greenaway & Haynes, 2003). Other studies suggest that the outcomes have indeed been negative, causing increased anxiety to students and also possibly affecting how they perceive the University-level study e.g. different options, more emphasis on value for money etc. (Declercq & Verboven, 2015; Wilkins et al., 2013). Galindo-Rueda et al. (2004:86) argued that 'our detailed individual level analysis suggests that in 1996, i.e. before tuition fees, there was certainly substantial social class educational inequality in HE but that it occurred largely as a result of inequalities earlier in the education system on HE participation, even after conditioning for the number of GCSEs and A levels that an individual has. This seems to suggest a widening of the social class gap in higher education itself in the period after the introduction of tuition fees'. This would suggest two key themes, firstly, that the introduction of fees did not serve to address the social inequalities both in terms of accessing and succeeding in HE. Secondly, that the problem might be in fact be impacted as well in the earlier stages of the education cycle. This would also echo with the earlier arguments discussed about the influence of wider contextual factors that span across the different layers of the policy lifecycle (Kim & Nuñez, 2013).

While arguably more selective funding to support students that from less affluent privileged backgrounds would certainly have merits, the picture is quite inconclusive as to how that is enacted in practice. This is can also be evidenced by a demonstrable shift on how the tuition fees are portrayed in education policy document. The most recent commissioned report to review post-18 education and finding published in May 2019 recommended the reduction of fees to 7,500, with the continued emphasis on deploying resources to disadvantaged students (Augar, 2019). While the suggestion to continue targeting students in need with financial aid is in-line with the previous mantra, the report does also suggest to 'bear down' on what they term as a significant minority of students enrolled on low-value degrees, leading to disappointment. Although the report does indicate in the main the value of HE remains high, universities should address this subpar provision and instead focus more on what the economy's needs are. This would possibly indicate a shift from the earlier push to continue increasing the sector continuously as with the increased supply, the environment is becoming saturated and the quality might suffer as a result (Charlton & Andras, 2002).

The drive to increase tuition fees since their introduction in 1998 had been common until now in terms of the result, despite differing parties being in power – firstly, the Labour and then the Conservatives. It would be interesting to turn our attention to the political dimension underpinning not only tuition fees, but more widely the policy around WP. Characteristically, even during the early days when government decided to expand the HE sector, although addressing inequalities was part of the agenda, alongside the social justice arguments there were voices about the need to satisfy the economy and its needs, by transforming universities to match (Carpentier, 2018). This was also coupled by a continued lack of funding that started in the 1980s and through the 1990s despite the policies to increase both the number of institutions and students going through university (Johnstone, 2005). Although part of the rhetoric behind the impetus to restructure and effectively restrict public subsidy for universities was also seen to provide greater support for the students in need, there are many voices that see a clear influence of neo-liberal economic trends that have led the evolution in the every strand of HE policy since 1990s (Johnstone, 2005; Lunt, 2008). Nevertheless, authors also do suggest that this historically went opposite to the traditional ideology of the labour party, but the broader developments to an extent had 'tied' the hands of the Blair's government in situation where 'economic and financial imperatives implied the creation of incentives for alternative (to the state) modes of funding higher education, and further drives for quasi-privatisation and entrepreneurial activity. In addition, the vision included a firm commitment to WP and greater social inclusion. But there remained the big political question of who should pay for continued expansion of higher education' (Lunt, 2008:744). This particular paradox, of having to address chronic financial deficiencies in relation to HE funding but still adhere to the tenets of breaking down social class barriers and enable mobility of under-represented groups through universities, seems to be colouring and driving the education policy, not only within the sphere of WP, but across the entire gamut of strategies and actions for education (Hill, 2010; Levidow, 2002; Naidoo & Williams, 2015).

Consequently, the drive towards marketisation of the HE sector in the UK could not leave the WP unaffected. McCaig (2015) argues that with the coalition government adopted a more selective approach to facilitating access to HE, moving from providing opportunities to all to focusing more on the 'bright' individuals from disadvantaged background and particularly towards more 'eclectic' universities. While this shift in policy could be partly attributed to the mounting pressures from the earlier interventions in HE, it was also further exacerbated by the global financial crisis of 2007/08, showing again how the macro-environmental contextual influences can well impact on policy formulation. The other interesting aspect of this reconfigured policy lies with the choice to target more elite institutions. This divide can be traced back to the University expansion in the early 1990s, when the promotion of former Polytechnics to universities created different groups, the so-called old universities, or Russell group and the newer post-1992s, with the former placing more emphasis on research and the latter on teaching (Evans et al., 2019).

4.5   DISCOURSES ON WP: HOW THE ACTORS INTERPLAY

The logic was to attempt to offer universities the opportunity to charge variable fees, depending on their reputation and overall quality enabling therefore better-performing institutions to impose higher fees if they so choose (Miller, 2010). Nevertheless, this was again rife by the spirit and principles dictated by economic rationalization and marketization of the University education (McCaig, 2015; Miller, 2010). Despite the theory and the drive towards education, the persisting inequality to-date would suggest that historically, there seems to be more focus on exogenous factors, particularly market and economy rules rather than the principles of social justice and enabling fairer access. Exploring the impact of Naidoo and Williams (2015) posited that institutions have been forced to concentrate on hard metrics that are tied to their position in the league tables as that determines funding, on the expense of students from less advantaged backgrounds, in order to avoid school leavers with lower scores that may have an adverse knock-on effect on the performance indicators universities have met. Even more, despite the obligation by the government for all HEIs to engage in WP activities (McCaig, 2015), there are concerns as to how effective these outreach activities are as 'the drive to "measure the measurable" may be undermining successful activities, while unhelpful inter-institution competition has replaced the co-operative ethos and wider social justice aims that dominated ten years ago' (Harrison & Weller, 2017:141). The competition among universities also raises another significant aspect as to how education policy in this area has been implemented. Institutions play a major role on how they approach the WP requirements, mainly due to their autonomy, but also because of regional differences across the different nations and areas of the UK (Gallacher & Raffe, 2012). This is also very heavily influenced by the vague conceptualization of what WP should actually entail and the differing priorities consecutive governments have pursued, despite the common ground around commodification of HE (Stevenson et al., 2010).

Jones and Thomas (2005) suggested that WP policy developments could be explored through the lenses of three models, utilitarian, academic and transformative. The academic model views the lack of participation as a consequence of attitudinal characteristics, with people from under-represented or disadvantaged socioeconomic groups naturally displaying a lack of predisposition towards HE. Therefore, raising expectations for the more gifted would provide an appropriate remedy to increasing participation. The model is often criticized for ignoring the root causes of the problems and the actual barriers students coming from less privileged strata face (Jones & Thomas, 2005; Sheeran et al., 2007). The utilitarian model encompasses the premise of 'low aspirations' but also recognises the possibility the lack of academic credentials as an additional obstacle to WP (Jones & Thomas, 2005). The institutional response therefore is to provide additional support through target curriculum and modules, to compensate for the academic shortcomings and facilitate the transition to HE (Shaw et al., 2007). Lastly, the transformative model espouses a more progressive approach, necessitating an overhaul of universities philosophy, culture and systems to reflect the actual needs of a truly responsive and reflexive participatory policy to HE (Jones & Thomas, 2005).

Although all three discourses have acted to shape and influence the formulation of education policy on WP, they have also exacerbated the complexity of the field, perpetuating the lack of clarity and direction in relation to the operationalization of said practices, often results in disparity on how institutions engage with WP and outreach initiatives (Burke, 2009; Jones & Thomas, 2005). Tangibly, this confusion can create significant practical constrains on the ground, with individual institutions and staff offer faced with confusion and inability to sustain a meaningful WP strategy and practices (Butcher et al., 2012; Stevenson et al., 2010). What is worthwhile to observe is the interplay across the various actors from political parties, regulatory bodies to institutions of HE. With regards the latter, although the element of autonomy is still strong, underpinned by competitive pressures and a more demanding environment, the bearing of macro-level influences is still considerably potent albeit in creating a more confounding and complex field.

4.6 CONCLUSION

Overall, trying to crystallise the main arguments analysed in this narrative, allows us to draw a few important conclusions. Firstly, WP continues to be at the forefront of policymakers' attention and features prominently in the government's education policy documents (e.g. see Augar, 2019). Nevertheless, despite the importance attributed by all actors and some positive empirical data, there are still concerns that there has been little actual progress in terms of eliminating barriers and enabling students from less privileged socioeconomic groups to attend university (Boliver, 2013; Galindo-Rueda et al., 2004). This could be due to a number of factors; the major interventions by increasing university numbers, positions, and changing the funding system with the imposition of fees, appear to be more driven by external stimuli from the market and economy instead of the requirements of the people from these groups. Equally, it is quite possible the lack of participation might be starting earlier at the education lifecycle and therefore interventions at the University level might be rendered considerably less effective (Galindo-Rueda et al., 2004). Another theme that emerged is the obvious interplay between the various actors and stakeholders and how the lack of clarity and ambiguity that permeates the discussions around this field has influenced not only policymaking, but also enactment (Kettley, 2007; Stevenson et al., 2010). Moving forward perhaps would require a reconsideration of how WP is defined, explored and articulated in a more consistent and systematic fashion, allowing the construction of a new architecture that would underwrite education policy in this area, with the principles that the Robins report espoused all those years back.

REFERENCES

Augar, P. (2019). Independent Panel Report to the Review of the Post-18 Education and Funding. Available via https://assets.publishing.service.gov.uk/government/uploads/system/uploads/attachment_data/file/805127/Review_of_post_18_education_and_funding.pdf (30/6/2019).

BIS (2014). *National Strategy for Access and Student Success*. London: BIS.

Boliver, V. (2013). How fair is access to more prestigious UK Universities?. *The British Journal of Sociology, 64*(2), 344–364.

Brown, R., & Carasso, H. (2013). *Everything for Sale? The Marketisation of UK Higher Education*. London: Routledge.

Burke, P.J. (2009, April). Widening participation: Identity, difference and in/equality. In *Keynote speech, aligning participants, policy and pedagogy: Tractions and tensions in VET research, 12th conference of the Australian VET Research Association, Crowne Plaza, Coogee, Sydney* (pp. 16–17).

Burke, P.J. (2012). *The Right to Higher Education*. Abingdon: Routledge.

Butcher, J., Corfield, R., & Rose-Adams, J. (2012). Contextualised approaches to widening participation: A comparative case study of two UK Universities. *Widening Participation and Lifelong Learning, 13*(1), 51–70.

Callender, C. (2013). Part-time undergraduate student funding and financial support. In C. Callender & P. Scott (Eds.), *Browne and Beyond: Modernizing English Higher Education*. London: IOE Press.

Carpentier, V. (2018). *Expansion and Differentiation in Higher Education: The Historical Trajectories of the UK, the USA and France* (p. 33). Centre for Global Higher Education Working Papers.

Charlton, B.G., & Andras, P. (2002). Auditing as a tool of public policy: The misuse of quality assurance techniques in the UK university expansion. *European Political Science, 2*(1), 24–35.

Clarke, P. (2018). Who Do You Know: The Importance of Social Capital in Widening Participation in HEPI Report N. 98. Available via https://www.hepi.ac.uk/wp-content/uploads/2017/08/FINAL-WEB_HEPI-Widening-Participation-Report-98.pdf (30/6/2019).

Declercq, K., & Verboven, F. (2015). Socio-economic status and enrollment in higher education: Do costs matter?. *Education Economics, 23*(5), 532–556.

Department for Education Skills (DfES) (2003). *Widening Participation in Education*. London: HMSO.

Evans, C., Rees, G., Taylor, C., & Wright, C. (2019). 'Widening access' to higher education: The reproduction of university hierarchies through policy enactment. *Journal of Education Policy, 34*(1), 101–116.

Galindo-Rueda, F., Marcenaro-Gutierrez, O., & Vignoles, A. (2004). The widening socio-economic gap in UK higher education. *National Institute Economic Review, 190*(1), 75–88.

Gallacher, J., & Raffe, D. (2012). Higher education policy in post-devolution UK: More convergence than divergence?. *Journal of Education Policy, 27*(4), 467–490.

Greenaway, D., & Haynes, M. (2003). Funding higher education in the UK: The role of fees and loans. *The Economic Journal, 113*(485), F150–F166.

Harrison, N., & Waller, R. (2017). Success and impact in widening participation policy: What works and how do we know?. *Higher Education Policy, 30*(2), 141–160.

HEFCE (2001). *Strategies for Learning and Teaching in Higher Education*.

Hill, D. (2010). Class, capital, and education in this neoliberal and neoconservative period. In *Revolutionizing Pedagogy* (pp. 119–143). New York, NY: Palgrave Macmillan.

Hubble, S., & Bolton, P. (2018). Higher Education Tuition Fees in England, House of Commons Library. Available via https://researchbriefings.parliament.uk/Research-Briefing/Summary/CBP-8151#fullreport (30/6/2019).

Johnson, J. (2015). Lifting the Cap on Student Numbers Will Drive Social Mobility Evening Standard. Available via https://www.standard.co.uk/comment/comment/jo-johnson-lifting-the-cap-on-student-numbers-will-drive-social-mobility-a2922356.html (30/6/2019).

Johnstone, D.B. (2005). Fear and loathing of tuition fees: An American perspective on Higher Education finance in the UK. *Perspectives: Policy and Practice in Higher Education, 9.1*, 12–16.

Jones, R., & Thomas, L. (2005). The 2003 UK Government Higher Education White Paper: A critical assessment of its implications for the access and widening participation agenda. *Journal of Education Policy, 20*(5), 615–630.

Kettley, N. (2007). The past, present and future of widening participation research. *British Journal of Sociology of Education, 28*(3), 333–347.

Kim, D., & Nuñez, A.M. (2013). Diversity, situated social contexts, and college enrollment: Multilevel modeling to examine student, high school, and state influences. *Journal of Diversity in Higher Education, 6*(2), 84.

Levidow, L. (2002). *Marketizing Higher Education: Neoliberal Strategies and Counter-Strategies* (pp. 227–248). Oxford: The Virtual University.

Lunt, I. (2008). Beyond tuition fees? The legacy of Blair's government to higher education. *Oxford Review of Education, 34*(6), 741–752.

Machin, S. & Vignoles, A., (2006). *Education Policy in the UK. CEE DP 57.* Centre for the Economics of Education. London School of Economics and Political Science, Houghton Street, London, UK.

Macleod, D. (1996, 12 March). Crusader takes up Cinderella case. *The Guardian*, 2.

Martin, I., (2018). Benchmarking widening participation: How should we measure and report progress. *Higher Education Policy Institute Policy Note*, 6.

McCaig, C. (2015). The impact of the changing English higher education marketplace on widening participation and fair access: Evidence from a discourse analysis of access agreements. *Widening Participation and Lifelong Learning, 17*(1), 5–22.

McCaig, C. (2018). English higher education: Widening participation and the historical context for system differentiation. In *Equality and Differentiation in Marketised Higher Education* (pp. 51–72). Cham: Palgrave Macmillan.

Miller, B. (2010). The price of higher education: How rational is British tuition fee policy?. *Journal of Higher Education Policy and Management, 32*(1), 85–95.

Murphy, R., Scott-Clayton, J., & Wyness, G. (2017). Lessons from the end of free college in England. In *Brookings Evidence Speaks Reports*.

Naidoo, R., & Williams, J. (2015). The neoliberal regime in English higher education: Charters, consumers and the erosion of the public good. *Critical Studies in Education, 56*(2), 208–223.

Pennell, H., & West, A. (2005). The impact of increased fees on participation in higher education in England. *Higher Education Quarterly, 59*(2), 127–137.

Riddell, S. (2015). *Higher Education in Scotland and the UK.* Edinburgh: Edinburgh University Press.

Shaw, J., Brain, K., Bridger, K., Foreman, J., & Reid, I. (2007). Embedding widening participation and promoting student diversity. In *What Can Be Learned from a Business Case Approach.* York, UK: Higher Education Academy.

Sheeran, Y., Brown, B.J., & Baker, S. (2007). Conflicting philosophies of inclusion: The contestation of knowledge in widening participation. *London Review of Education, 5*(3), 249–263.

Stevens, R. (2004). *University to Uni.* London: Politico's Publishing.

Stevenson, J., Clegg, S., & Lefever, R. (2010). The discourse of widening participation and its critics: An institutional case study. *London Review of Education, 8*(2), 105–115.

The Committee on Higher Education (1963). Higher Education Report of the Committee Appointed by the Prime Minister under the Chairmanship of Lord Robbins, London. Available via http://www.educationengland.org.uk/documents/robbins/robbins1963.html (30/6/2019).

Trowler, P.R. (2002). *Higher Education Policy and Institutional Change. Intentions and Outcomes in Turbulent Environments.* Buckingham: Open University Press.

Wilkins, S., Shams, F., & Huisman, J. (2013). The decision-making and changing behavioural dynamics of potential higher education students: The impacts of increasing tuition fees in England. *Educational Studies, 39*(2), 125–141.

5 The Role of Public Universities in Entrepreneurship in Mexico

María del Rocio Soto Flores, Ingrid Yadibel
Cuevas Zuñiga, and Ericka Molina Ramírez
Instituto Politécnico Nacional, Juárez, Mexico

CONTENTS

5.1   INTRODUCTION: BACKGROUND AND DRIVING FORCES

Currently, entrepreneurship is one of the most important variables for the economic development of a country. The creation of new companies in traditional activities and also those that are totally innovative can generate great benefits to the entrepreneur of the new business and to the national economy. This beneficial effect expands in a positive way, benefiting families, society and countries in general.

One of the great problems in the world's economies is their lack of capacity to generate the jobs that a society in constant growth demands. Where youth unemployment has been one of the hardest hit strata, for example, in the euro zone (OECD/EU, 2017), "it reached a maximum of 24% in 2013 at the level of the European Union and more than 50% in some member states". While, in Mexico, the National Institute of Statistics and Geography reported that unemployment in 2019 in those under 25 years of age was 6.9% of the total economically active population and in 2020, derived from the pandemic caused by the SARS-CoV-2, unemployment in that age stratum increased to 8.1%.

DOI: 10.1201/9781003021230-5

Against this background, governments and universities are aware of the importance of promoting entrepreneurship, and for this reason, they have placed great interest in its study and in the formulation of strategies to promote it. In this vein, the Lederman et al. (2014) indicates that exercises aimed at identifying and analysing the technological behaviour of innovation and entrepreneurship agents, measuring their innovative efforts and evaluating the results achieved should be thought of as tools of strategic importance to guide public and private actions.

In addition to the above, OECD/EU (2017), inclusive entrepreneurship policies can play an important role in addressing these issues by offering people the opportunity to make an economic and social contribution. These policies and programmes can benefit people by enabling them to learn skills, develop networks and generate their own income, whether by starting a business or by acquiring the skills and experience that will help them find employment. They also provide an opportunity for economies to grow through the economic contribution of unnecessary or underutilized resources.

5.2   THE EMERGENCE OF ENTREPRENEURSHIP

According to the report by The Kauffman Foundation (2007), entrepreneurship is the transformation of an innovation into a sustainable company that generates value. For Braunerhjelm (2010), "all are entrepreneurs only when new combinations of techniques and activities are carried out, and that characteristic is lost as soon as a business is created, stabilizes and begins to manage as anyone else would manage their business". Lederman et al. (2014) refer that behind the most dynamic and productive companies – those that innovate, whose production is expanding and rate of job creation is relatively high – there are creative entrepreneurs.

Entrepreneurs innovate new ways of manipulating nature and new ways of gathering and coordinating people [...] the innovator shows that a product, a process or a mode of organization can be efficient and profitable (The Kauffman Foundation, 2007), and this has positive effects on the entire economy from the point of view of job creation, development of new products and services, new market opportunities and national and regional economic growth. The Global Entrepreneurship Monitor (GEM), in its Global Report (2010), points out that "entrepreneurs drive and shape innovation, accelerating structural changes in the economy. By introducing new competition, they indirectly contribute to productivity. Entrepreneurship is therefore a catalyst for economic growth and national competitiveness".

For the World Economic Forum, entrepreneurship is based on pillars that form its own ecosystem for taking advantage of opportunities, within the main accessible markets, human capital and labour force and financing (World Economic Forum, 2014). However, to develop and increase the characteristics of entrepreneurship and the entrepreneur, the institutions that manage knowledge, promote an efficient macroeconomic framework, and encourage the attitudes of individuals (entrepreneurial spirit) are part of the entrepreneurial scenario (Braunerhjelm, 2010).

However, regarding entrepreneurship in Latin America and according to data from ICSEd-Prodem (2018), it is very uneven in the region. The five best-evaluated countries in Latin American entrepreneurship are: (1) Chile, (2) Argentina, (3) Mexico, (4) Costa Rica and (5) Colombia. The largest deficits in Latin American countries are in the science, technology, and innovation (STI) platform, in the business structure, access to social capital and financing. This scenario limits the options for value propositions to flourish based on the results of scientific research and its articulation with companies and on their own demands.

5.3 UNIVERSITIES AND THEIR RELATIONSHIP WITH ENTREPRENEURSHIP

For years, the universities of the world evolve to adapt to the changes demanded by rapid technological change, the knowledge society and innovative entrepreneurship, which leads to a modification in organizational structures and the redesign of the traditional education model towards another based on competencies and intensive use of technology in teaching-learning processes, more in line with the changes in the environment and demands of digital societies.

Universities maintain a responsibility to society, so they need to: (1) work on long-term goals by promoting critical thinking through teaching and research, and demonstrate respect for diversity; (2) communicate the crucial role of research based on the autonomy of universities by guaranteeing academic freedom, which contributes to high quality teaching, improving skills and their commitment to the transmission and dissemination of knowledge; (3) consolidate alliances with other stakeholders; (4) strengthen ties at the regional and local level, since universities train society (European University Association, 2003) to enter the labour market, in training for entrepreneurship and the foundation of a global information society.

In this sense, it is irrefutable that educational institutions such as universities play an essential role in the development of a society, forming quality intellectual capital through adequate teaching to students, who are a key factor in the production of knowledge. In addition, it is a motivational catalyst for students to acquire this knowledge and consequently increase the indices of entrepreneurship based on knowledge, creativity and talent for the generation of innovative companies that contribute to the growth of the economy and society as a whole.

In the current climate of innovation agglomeration, some public sector institutions – especially universities – are reinforced by market forces, which make some of them more attractive to students, teachers and funders (WIPO, 2019).

The capacity of a country to progress and prosper in a global environment based on the rapid generation and application of knowledge in economic activity depends on its capacities to educate its entire population with quality. In this challenge, universities play a leading role in the formation of highly qualified human capital for entrepreneurship and innovation.

This is the case of entrepreneurship and innovation programmes in countries such as the United States (Vicens & Grullón, 2011), where universities are a fundamental part of the most prominent entrepreneurship and innovation ecosystems and serve as a benchmark throughout the world. For example, Stanford, Babson and Georgia Tech universities are successful cases in areas with different demographic and cultural conditions.

That is, the responsibility of creating new knowledge, training and providing experience to the entities of society and managing knowledge (intellectual capital and technology), falls on educational institutions (Adhikari, 2010), and "the knowledge acquired creates continuous innovation and allows the development of competitive advantages" (Carreon & Melgoza, 2012). These aspects are important for entrepreneurship, as they involve new ways of doing business, generating new products and new ways of solving a problem, where experiences learned and information received from universities play a leading role. But, in addition, the creation of companies creates momentum (The Kauffman Foundation, 2007) and "generates continuous innovation and improves products, services and institutions".

Unfortunately, the levels of innovation in Latin America and the Caribbean still lag in this regard. The Economic Commission for Latin America and the Caribbean published that this lag is due to the lack of technology because of the economic structure of the region but emphasizes that there are other factors that limit innovation such as the low participation and performance of academic institutions (Rivas & Sebastian, 2014). However, it should be noted that, in these economies the TEA (business activity rate) showed a dynamic and high behaviour in the age group 18–24 (16.5%), 35–44 (20.6%) and 45–54 years (17.9%), according to data from the GEM (GEM, 2017/2018).

Although it is also true that "the relationship between educational level and the propensity to start a new business is complex, especially because of the positive association between education and lifetime income, and any link between education and the ability to detect opportunities" (GEM, 2020/2021), and this is where the quality and educational level play a key role in generating creative ideas, developing capacities and skills for entrepreneurship.

However, to move towards an economy that generates employment and national wealth, universities must extend their substantial functions to other strategic areas, focused on promoting entrepreneurship in teaching and motivating students to create their own businesses in which they apply the knowledge acquired throughout their training. Entrepreneurs "consider three pillars as the most important for achieving growth: accessible markets, human capital and workforce, and financing" (World Economic Forum, 2014). For example, for the United States of America, the generation of new knowledge is the highest expression of learning; therefore, entrepreneurship and university education are inextricably linked to each other (The Kauffman Foundation, 2007).

The education, knowledge and skills acquired in universities are decisive elements for the future of young professionals and entrepreneurs, who face new demands in labour markets marked by rapid technological advances, digital transformation and economic globalization.

In this context, "knowledge and the institutions that govern the dissemination and maintenance of knowledge seem to be the most representative aspect of innovative entrepreneurship. Individuals with certain abilities and characteristics tend to engage in an entrepreneurial process characterized by research, uncertainty, and randomness. One characteristic that stands out is the involvement of individuals in constant experiments, which allow trying different ideas and models before finding the right one" (Braunerhjelm, 2010) and according to the new challenges of a changing world, manifested by information societies and digital businesses.

Organization for Economic Cooperation and Development (OECD) data emphasize that around four out of ten adults with a higher education (university) degree complete on-the-job training or internships, while around eight out of ten with low levels of education do not. Low-skilled workers are also less likely to participate in these trainings, while they are more likely to work in sectors that evolve with automation and digital development, that is, to have jobs in which they will need to acquire new skills in the short and long terms. (OECD, 2021). In this sense, permanent training, developing and promoting an entrepreneurial culture in young people, the stratum most affected by the high levels of unemployment, in the case of Mexico, in 2019, unemployment in those under 25 reached levels of 6.9% of the economically active population, while, in 2020, unemployment in the same age range amounted to 8.1%, a situation that coincides with the crisis generated by the COVID-19 pandemic. Therefore, starting your own business is a feasible option for young people and people in general, to develop in a changing environment and eminent uncertainty.

5.4 MEXICAN PUBLIC UNIVERSITIES AND ENTREPRENEURSHIP

It should be noted that developing countries, including Mexico, although they are still far behind the countries of the OECD in terms of their ability to introduce innovations in markets, have registered in the last decade a significant increase in its business dynamism, within the framework of a renewed interest in the generation and dissemination of knowledge, and the emergence of innovative entrepreneurs with high growth potential (OECD, 2015) are some of the strategies that Mexico has followed, to disseminate knowledge in support of economic growth and the generation of new companies that contribute to national wealth and employment.

Adhikari (2010) refers that "the responsibility of creating new knowledge, training and providing experience to the entities of society, as well as managing knowledge (intellectual capital and technology) falls on Higher Education Institutions", and Mexico has an educational performance below OECD member countries and a lower rate of graduates from higher education institutions (HEIs) than Argentina (40%), Brazil (57%) and Chile (35%) (Kuznetsov & Dahlman, 2008). And when other indicators such as research and development (R&D) are involved, enormous differences are identified; for example, the government of Mexico only invests 0.5% of its GDP in R&D, while economies such as Brazil have been investing 1.0% of its GDP in recent years in R&D. These efforts have yielded positive results for the Brazilian economy, because according to WIPO (2019), "it is the only country in the region [Latin America] that has science and technology centres among the top 100 in the world".

For Stam and Spigen (2016), "the generation of entrepreneurship in universities must have as a goal a productive entrepreneurship that has a real impact on society, the economy, and the natural environment of a country, through its students (future entrepreneurs) who explore new opportunities and add as much value as possible to new products and services". In an entrepreneurial ecosystem, universities are part of the fundamental institutions that reinforce human capital and entrepreneurial culture, and in turn, the latter two support the growth and development of universities. However, what is important are the students and their creativity to contribute new ideas and increase the intellectual capacity of a community (Manson & Ross, 2013).

Although universities do not act alone to promote entrepreneurship and knowledge, since they do so in collaboration with the government and companies, it is in universities where human capital is found with the knowledge and skills to make use of their creativity to start a new business. It is worth mentioning that Mexico is made up of 32 states, and in each of them, there is at least one public university, except for Mexico City, where three of the largest and most important universities in the country are located.

Among these three, the programmes implemented by the National Autonomous University of Mexico (UNAM) and the National Polytechnic Institute (IPN), aimed at promoting entrepreneurship, stand out. In the case of UNAM, it does so through its Business incubator and the entrepreneurs and innovation programme of the Coordination of Innovation and Development (CID). Regarding the IPN, it promotes and supports entrepreneurship through its Center for Incubation of Technology-Based Companies (CIEBT) and the Poliemprende programme, in addition to having an office in collaboration with the Mexican Institute of Industrial Property (IMPI), to provide advice and support for the protection of the polytechnic inventiveness of students and researchers.

For universities, it represents an added value to develop spaces, programmes and workshops that allow interaction between different agents capable of creating knowledge, as well as the exchange of ideas, techniques and experiences to obtain a new or different result. And in the same way that one interacts internally within an institution, interactions with all those external stakeholders represent an opportunity for entrepreneurship (Mendoza et al., 2019). The entrepreneurial ecosystem matures according to the dynamism of a country's economy, its universities, public policies, infrastructure, markets and those services that serve as a basis for the application of creativity (Stam & Spigen, 2016). Developing an entrepreneurial culture will give value and drive to the entrepreneurial ecosystem, and its creation and maturity depend on the institutions that generate and share knowledge.

Universities have the power to motivate students to question the functioning of a social, economic and cultural system, and to contribute ideas and test their results through research and experimentation. The fear of risk is a limitation of Mexican society. In Mexico, only 35 people out of every 10,000 people in the EAP (economically active population) apply a financial culture, which is a very low figure compared to the 60 people out of 100 in the United States (CONDUSEF, 2018). It is here where public policies and universities have to focus their efforts to promote a culture of innovation and entrepreneurship based on education, knowledge, skills, managerial instructions, confidence and attitude for entrepreneurship.

One of the great failures that inhibit entrepreneurship and fear of risk in the working-age population is related to the fact that in Mexico (INEGI, 2020), eight out of ten companies fail before reaching their first 5 years of life, and approximately 75% of the enterprises that are founded as small and medium-sized enterprises (SMEs), do not reach 10 years of age, despite the fact that the country is considered a suitable place for entrepreneurship. To these data, adds Velasco (2019), 80% of entrepreneurs sell less than one million pesos per year; 89% of projects start with personal or family financing against 3% that receive formal investment; only 14% of start-ups receive income from international markets and 52% of these businesses are run from home or do not have a formal physical office.

Although the health contingency caused by COVID-19 has been a determining factor in the closure of many companies, entrepreneurship statistics in Mexico have not changed trend in recent years, mentions INEGI.

However, the referred data, it is important to take advantage of the rise in young Mexicans who want to start a new business, since it is estimated that (INEGI, 2020), "46% of young people between 34 and 19 years old see a feasible scenario to start an entrepreneurial activity, in addition to verifying that 60% are considered suitable to undertake", which allows inferring that, subject to the economic changes derived from the COVID-19 pandemic, entrepreneurship in Mexico is estimated as a perspective of fighting unemployment and poverty.

Given the health contingency and the economic effects it has had on the entire national economic activity, INEGI (2020) reports that, although there is no concrete data on entrepreneurship in Mexico, it is considered that among Mexican entrepreneurs, 40% are mainly young people between 22 and 34 years old, which allows us to infer that the level of young university students is the one that produces the most entrepreneurs, but most of them start without a proper fiscal activity (registration), which makes their measurement difficult.

In the publication of the Global Innovation Index (2020), it is highlighted that in "Mexico there is a lot of talent and creativity, but it is the organizations that have the opportunity and resources to take advantage of them and generate the greatest innovation. [It is mentioned that] Mexico with a rating of 33.6 out of 100, rises one position compared to 2019, from place 56 to 55 on the global scoreboard and from third to second in Latin America. In Latin America, Mexico is the world's largest exporter of creative products".

It should be noted that in recent years there have been many development and promotion programmes for entrepreneurship in Latin America and Mexico (Vicens & Grullón, 2011), which include training on issues related to the area – mainly in universities – and platforms of support, such as incubators, networks of mentors and angel investors, financing programmes, entrepreneurship colloquia and awards for the best entrepreneur. Entrepreneur training programmes are focused on preparing a business plan once the entrepreneur identifies an idea. The most prominent ideas become part of incubators and mentor networks.

In fact, the GEM (2019–2020) highlights for the case of Mexico and its National Entrepreneurship Context Index (NECI), noting that Mexico ranks fourth globally in entrepreneurship education, despite not promoting it in basic levels. Similarly, Mexico is well positioned in social and cultural norms; this means that there are many people who see entrepreneurship as an option for a career. The promotion of entrepreneurship in basic education is a pending of Mexico and adds to the recommendations of international organizations such as the OECD, which suggests improving basic levels of education and all levels of training.

In other words, entrepreneurship in Mexico continues to be a great challenge for universities and governments. Universities should participate in training in a more dynamic way, since many of the entrepreneurs who run a micro, small and medium-sized company have little training in abilities management, innovation and marketing skills for the sustainability and growth of their business. In fact, training programmes in business administration, economics and engineering have weaknesses in promoting innovation, creativity and self-employment, which are essential in entrepreneurship instruction.

In this sense, the opinion of The Kauffman Foundation (2007) is categorical in stating that "human understanding, ingenuity and inventiveness will be increasingly critical to create a sustainable future. But innovation alone will not be enough. We will need people who know how to implement new ideas and make them accessible to large populations. An entrepreneurial society will not arise or persist by accident. We will have to build and maintain it. To do both, we will have to understand why entrepreneurship is important, how it works, and how to sustain it. That understanding is the result of education".

In other words, entrepreneurship must be promoted through training and solid training to undertake, innovate and compete in a global economy. Therefore, universities have a preponderant role and a social responsibility as the main generators of knowledge and training of human capital capable of generating creative ideas and skills for the creation of new companies that achieve a multiplier effect in the national economy. Therefore, it is important to outline some recommendations that support entrepreneurship in Mexico.

5.5 SOME RECOMMENDATIONS TO PROMOTE ENTREPRENEURSHIP FROM PUBLIC UNIVERSITIES IN MEXICO

A first recommendation is "to turn universities and laboratories into centres where students develop their creativity, their capacity for innovation and their way of thinking through practical creativity exercises" (Seelig, 2009, cited by Vicens and Grullón, 2011).

It is essential that Mexican universities have a solid organization at their different levels of action in terms of efficiency, quality and excellence. For this reason, it is vital that they review their organizational structure and substantive functions, to orient them towards an education more in line with global economies, digital societies, disruptive technological change and an entrepreneurial spirit to promote national entrepreneurship.

The government, in addition to promoting and strengthening the entrepreneurial ecosystem, requires greater commitment in generating the infrastructure and agencies to support entrepreneurship, as well as establishing financing programmes and instruments necessary for this purpose. Increase investment in R&D to influence the creation of new knowledge and innovation within universities and public research centres, these being, in the Mexican case, the main generators of knowledge and innovation. In the developed world, governments "are investing in national prosperity by promoting lucrative new technologies in conjunction with universities to serve as an economic, social, and cultural engine in the regions" (Välimaa & Hoffman, 2008).

Mexican public universities must "work on long-term goals to strengthen critical reasoning through teaching and research" (European University Association, 2003) and influence the development of entrepreneurship and a competitive economy based on knowledge, capable to achieve economic progress for social mobility, the creation of better paid jobs and promoting university entrepreneurship.

Mexican public universities must be more dynamic in establishing strategic alliances with other national and foreign institutions for the production, exchange and dissemination of knowledge, which in turn is transferred to society, the economy and the entrepreneurial spirit and dynamism.

5.6 CONCLUSIONS

An entrepreneur is a creative individual with the ability to take risks and start a new business, to take advantage of market opportunities and launch a new product or service, to satisfy needs by combining new production methods and techniques and assuming a risk with eminent uncertainty.

Entrepreneurship in Latin America is very uneven for the region. The five best evaluated countries in Latin American entrepreneurship are Chile, Argentina, Mexico, Costa Rica and Colombia, while the rest of the countries present greater lags in the evaluation indices. The most important deficits are in the STI platform and in the business structure.

Universities play a key role in the training of highly qualified human resources demanded by the global economy. Within universities, many ideas, knowledge and innovation are generated, which is why the transfer of this knowledge should be used to support and encourage university entrepreneurship.

Recent studies identify that more and more new generations choose to start a new business and thereby consolidate an entrepreneurial culture and the creation of companies based on innovation in Mexico. However, 75% of the start-ups created expire before two years.

Although Mexican public universities have implemented various programmes aimed at instituting skills and entrepreneurial spirit in their students, due to the scarce information and availability of data from them, very little is known about the results achieved by the application of the university entrepreneurship programmes, making it difficult to obtain conclusive results at the national level.

REFERENCES

Adhikari, D. R. (2010). Knowledge management in academic institutions. International Journal of Educational Management, 24(2), 94–104.

Braunerhjelm, P. (2010). Entrepreneurship, Innovation and Economic Growth. Past experiences, current knowledge and policy implications. Stockholm: Swedish Entrepreneurship Forum and The Royal Institute of Technology.

Carreon, H. & Melgoza, R. (2012). México hacia una sociedad del conocimiento. Nóesis. Revista de Ciencias Sociales y Humanidades, 21(41), 122–135.

Condusef (2018, 10 10). Pierde el miedo. Retrieved from Condusef: https://www.condusef.gob.mx/Revista/index.php/inversion/otros/303-pierde-el-miedo

European University Association (2003). Response to the Communication from the Commission. The Role of the Universities in the Europe of Knowledge. European University Association.

Global Entrepreneurship Monitor (2010). Global Report 2010. London: GEM.

Global Entrepreneurship Monitor (2017). Policy Briefs 2017. Recuperado de: https://www. gemconsortium.org/

Global Entrepreneurship Monitor (2018). Global Report 2017/2018. Recuperado de: https:// www.gemconsortium.org/

Global Entrepreneurship Monitor (2019). Global Report 2018/2019. Recuperado de: https:// www.gemconsortium.org/

Global Entrepreneurship Monitor (2021). Global Report 2020/2021. London: Global Entrepreneurship Research Association. London Business School.

INEGI (2020). Datos de emprendimiento en México. México: INEGI.

Kuznetsov, Y., & Dahlman, C. (2008). Mexico's Transition to a Knowledge-Based Economy. Challenges and Opportunities. Washington, DC: The International Bank for Reconstruction and Development/The World Bank.

Lederman, D., et al. (2014). El emprendimiento en América Latina: muchas empresas y poca innovación. Washington, DC: The World Bank.

Manson, C., & Ross, B. (2013). Entrepreuneurial Ecosystems and Growth Oriented Entrepreuneurship. The Hague: OECD LEED Programme and the Dutch Ministry of Economic Affairs.

Mendoza, M., et al. (2019). The Knowledge Society and Entrepreneurship Within the Higher Education Institutions in Mexico. In E. Cardoso, The Formation of Intellectual Capital and Its Ability to Transform Higher Education Institutions and the Knowledge Society. Washington, DC: IGI Global.

OECD (2015). Start-up América Latina. Construyendo un futuro innovador. París: OECD.

OECD (2021). OECD Skills Outlook 2021. Learning for Life. Paris: OECD. (http://doi.org/ 10.1787/fc97e6d3-en).

OECD/EU (2017). Pallier la pénurie d'entrepreneurs 2017. Politiques de l'entrepreneuriat inclusif. Paris: OECD/EU.

Rivas, G., & Sebastian, R. (2014). Nuevas instituciones para la innovación: Prácticas y experiencias en América Latina. Santiago de Chile: CEPAL; Alis; Cooperación Alemana.

Seelig, T. (2009). What I wish I knew when I was 20: A crash course on making your place in the world. New York: HarperCollins Publishers.

Stam, E., & Spigen, B. (2016). Entrepreneurial Ecosystems. Utrecht: Utrecht School of Economics.

The Kauffman Foundation (2007). Entrepreneurship in America higher education. Kauffman: The Foundation of Entrepreneurship.

Välimaa, J., & Hoffman, D. (2008). Knowledge society discourse and higher education. Jyvaskyla: Springer.

Vicens, L. & Grullón, S. (2011). Innovación y emprendimiento: Un modelo basado en el desarrollo del emprendedor. BID/COMPETE CARIBBEAN.

WIPO (2019). World Intellectual Property Report 2019: The geography of innovation: Local hubs, global networks. Geneva: WIPO.

WIPO (2020). Global Innovation Index 2020: Who will finance innovation? Geneva: WIPO.

World Economic Forum. (2014). Entrepreneurial Ecosystems Around the Globe and Early-Stage Company Growth Dynamics. Geneva: World Economic Forum.

6 Working with Students on Establishing a Student-Oriented Classroom Culture

A Teaching Initiative Designed to Build an Inclusive and Highly Engaging Learning Environment in Online and Face to Face Environments

Biginas Konstantinos and Sindakis Stavros
Hellenic American University, School
of Business, Nashua, New Hampshire, USA

CONTENTS

6.1 INTRODUCTION

Student engagement has been linked to improved achievement, persistence and retention (Finn, Fiin, & Zimmer, 2021; Kuh, Cruce, Shoup, Kinzie, & Gonyea, 2008), with disengagement having a profound effect on student learning outcomes and cognitive

DOI: 10.1201/9781003021230-6

development (Ma, Han, Yang, & Cheng, 2015), and being a predictor of student dropout in both secondary school and higher education (Finn & Zimmer, 2012). Student engagement is a multifaceted and complex construct (Appleton, Christenson, & Furlong, 2008; Ben-Eliyahu, Moore, Dorph, & Schunn, 2018), which some have called a "meta-construct" (e.g. Fredricks, Blumenfeld, Friedel, and Paris, 2004; Kahu, 2013), and likened to blind men describing an elephant (Baron & Corbin, 2012; Eccles, 2016). There is ongoing disagreement about whether there are three components, e.g. (Eccles, 2016) – affective/emotional, cognitive and behavioural – or whether there are four, with the recent suggested addition of agentic engagement (Reeve, 2012; Reeve & Tseng, 2011) and social engagement (Fredricks, Filsecker, & Lawson, 2016). There has also been confusion as to whether the terms "engagement" and "motivation" can and should be used interchangeably (Reschly & Christenson, 2012), especially when used by policymakers and institutions (Eccles & Wang, 2012). However, the prevalent understanding across the literature is that motivation is an antecedent to engagement; it is the intent and unobservable force that energises behaviour (Lim, 2004; Reeve, 2012; Reschly & Christenson, 2012), whereas student engagement is energy and effort in action; an observable manifestation (Appleton, Christenson, & Furlong, 2008; Eccles & Wang, 2012; Kuh, 2009; Skinner & Pitzer, 2012), evidenced through a range of indicators.

Student engagement is the energy and effort that students employ within their learning community, observable via any number of behavioural, cognitive or affective indicators across a continuum. It is shaped by a range of structural and internal influences, including the complex interplay of relationships, learning activities and the learning environment. The more students are engaged and empowered within their learning community, the more likely they are to channel that energy back into their learning, leading to a range of short- and long-term outcomes, that can likewise further fuel engagement.

This conceptually driven set of elements, all of which make the "classroom culture", is seen as the building blocks within the learning community to allow for the creation of the most effective design for a learning situation based on the principles of mutual trust, empowerment, respect and responsibility.

Research Questions

1. How did the implementation of the classroom culture approach has impacted on student engagement levels?
2. Which indicators of cognitive, behavioural and emotional engagement were identified in the student population where the classroom culture approach has been implemented?
3. What were the most effective learning scenarios, modes of delivery and educational technology tools employed in the implementation of the classroom culture approach and generated positive results?

Research Objectives

1. To critically evaluate the implementation of the classroom culture approach on online and face to face environments.
2. To evaluate the significance of co-creating a commonly accepted set of rules and behaviours with students for the achievement of high levels of student engagement and involvement.

3. To examine the correlation between the implementation of the classroom culture approach and high levels of student performance and low absenteeism.
4. To critically discuss any potential variations in responses that are relevant to gender, age and cultural background.
5. To critically examine potential differences in the implementation and effectiveness of the classroom culture approach between face to face and online educational settings.

6.2   CLASSROOM CULTURE APPROACH – WHAT IS IT?

This is an approach that was first implemented during Jan–April Term 2019 for specific undergraduate and postgraduate modules. There is a continuation of that research effort with the implementation of the approach on online classes with diverse student populations. The "classroom culture" approach was accepted in the Teaching and Learning Conference – Advance HE Conference, 2020 and 2021.

Working with students, the researcher co-created the "classroom culture" in Week 1 and as a result everyone respected it. There was an exchange of views and agreement on a set of rules that were mutually agreed. That was a key success factor.

The classroom culture approach is based on the fundamental stakeholder management theory. Involvement and engagement of students in the decision-making process is essential to deal with any resistance and lack of interest throughout the term. Making students feel like active stakeholders (Jeffrey, 2009). A significant part of this approach is the use of education tools, which give students autonomy, encourage creativity, and facilitate communication and collaboration. Examples of digital education tools the researcher used consistently in class are Kahoot quizzes – often developed by the students themselves – Socrative and TED-ed – a platform that allows democratizing access to information for both teachers and students. Hence, people can have an active participation in the learning process of others. E-Learning scenarios are also a very effective for engaging learners and initiating positive behavioural change. Scenario-based learning is engaging, emotionally impactful and memorable. Workplace scenarios, cases and stories draw the learner in and prime them for the core content about ethical dilemmas and decision-making. The previous approaches are conducive to constructivist learning theory, a theory that stipulates that learners construct knowledge rather than just passively take in information. As people experience the world and reflect upon those experiences, they build their own representations and incorporate new information into their pre-existing knowledge (schemas).

Consequences of constructivist theory (Bada & Olusegun, 2015) are that:

• Students learn best when engaged in learning experiences rather passively receiving information.
• Learning is inherently a social process because it is embedded within a social context as students and teachers work together to build knowledge.
• Because knowledge cannot be directly imparted to students, the goal of teaching is to provide experiences that facilitate the construction of knowledge.

The common objective was to maximize students' learning experience through enjoyment, participation, learning, respect and punctuality. In such way, the students' feeling of belongingness (Maslow's hierarchy of needs) is enhanced and allows them to achieve very good results.

Apart from the standard university rules and expectations, the researcher introduced and discussed with students the concept of the classroom culture approach. There was a common view that the maximization of students' learning experience through commonly agreed accepted rules of learning was a common objective.

In other words, the researcher defined how things are being done around here (Deal & Kennedy, 2000) and discussed and agreed on acceptable/non acceptable behaviours.

A Few Examples

- Be in class (on campus and online) 5 minutes before the class begins. Students who are late need to wait until the first break before enter;
- Everyone, in every group, participates in class discussion and activities;
- Respect each other and do not interrupt when classmate talk; listen attentively to all students presentations (class activities) and ask questions in the end so meaningful discussions take place;
- Do not use electronic devices in class – except from research purposes;
- Provide flexibility on how to work on homework activities and present in class (presentations, role playing etc.);
- About 100% student led activities where the tutor was there to observe;
- Introduction of the "breaking news" activity where students bring in class relevant contemporary articles for discussion;
- Keep up with the time schedule – students knew, for example, that specific activities should start 5 minutes after the class begins. They were all ready and logged in on time in each and every class;
- Introduction of the "teaching assistant" scheme;
- Students to send their homework to the tutor's email the night before the seminar;
- Students to show drafts of their work for their summative assignments on pre-specified days during the term;
- Use of online learning tools and platforms to fully utilise educational technology.

6.3   METHODOLOGY

As this is an ongoing research project, a quantitative approach is being applied through the use of online questionnaires where students are invited to comment on their overall engagement and learning experience from the implementation of the classroom culture approach. Students have been gladly participating in such a research initiative. This is a result of a positive teaching and learning relationship that has been developed between the researcher and students. The usual University research ethics process is being followed and the ethical issues in the researcher

– participant relationship as identified by Saunders et al. will be addressed. Moreover, the six key principles for ethical research identified by ESRC will be honoured:

- research should aim to maximise benefit for individuals and society and minimise risk and harm
- the rights and dignity of individuals and groups should be respected
- wherever possible, participation should be voluntary and appropriately informed
- research should be conducted with integrity and transparency
- lines of responsibility and accountability should be clearly defined
- independence of research should be maintained and where conflicts of interest cannot be avoided they should be made explicit.

Qualitative oral feedback from students during and after the end of the teaching term is also being received, stored, and analysed. The feedback so far has been very positive.

Combining quantitative and qualitative data will allow:

Enriching: Using qualitative data to identify issues or obtain information on variables not found in quantitative surveys.

Explaining: Using qualitative data to better understand unexpected results from quantitative data.

Through observation while delivering modules and students' behaviour during the teaching term, attendance and marking records, the researcher witnessed that students benefited from their experience and enjoyed the class.

In addition, with the intent to systematically map empirical research on student engagement and educational technology in higher education, a systematic review will be conducted. A systematic review is an explicitly and systematically conducted literature review that answers a specific question through applying a replicable search strategy, with studies then included or excluded, based on explicit criteria (Gough, Oliver, & Thomas, 2012). Studies included for review are then coded and synthesized into findings that shine light on gaps, contradictions or inconsistencies in the literature, as well as providing guidance on applying findings in practice. A thorough comparison between primary data findings and existing literature will be conducted.

6.4   THE STUDENT ENGAGEMENT FRAMEWORK

There are three widely accepted dimensions of student engagement: emotional, cognitive and behavioural. Within each component there are several indicators of engagement. Engagement is more than involvement or participation – it requires feelings and sense making as well as activity (see Harper & Quaye, 2009, 5). Acting without feeling engaged is just involvement or even compliance; feeling engaged without acting is dissociation. Although focusing on engagement, Fredricks, Blumenfeld, Friedel, and Paris (2004, 62–63), drawing on Bloom (1956), usefully identify three dimensions to student engagement, as discussed in the following:

1. *Behavioural engagement*: Students who are behaviourally engaged would typically comply with behavioural norms, such as attendance and involvement, and would demonstrate the absence of disruptive or negative behaviour.
2. *Emotional engagement*: Students who engage emotionally would experience affective reactions such as interest, enjoyment, or a sense of belonging.
3. *Cognitive engagement*: Cognitively engaged students would be invested in their learning, would seek to go beyond the requirements, and would relish challenge.

A Few Evidence and Feedback

- Overall, the researcher has received so far enthusiastic verbal and written feedback from students.
- Students pass modules with very good results and a good percentage of distinctions.
- Good quality of written assignments and group presentations.
- Very high attendance records for classes.
- The overall satisfaction rates for the modules remain very high. (90–100%)
- Students invested in their learning, seek to go beyond the requirements, and relish challenge.

Students' Feedback Sample

- "Got my Result, Firstly, I want to Thank you for your support and I hope I will get support like this in future as well once again thank you"
- "We really like your class which give us freedom and encourage us to talk more, more critical thinking"
- "Just writing to say a HUGE thank you for conducting great seminar classes this term. Really enjoyed our time spent with you this term and looking forwards to hopefully having you as a supervisor for internship next term".
- "I did not want to miss even one class for this module".

6.5   PRACTICAL IMPLICATIONS

The seminars/webinars consisted of a mixture of student led seminars, scenario-based activities, self-discovery activities, ongoing reflection, an invitation of guest speakers from various fields and geographical locations and debates on contemporary leadership/management issues. Online International guest speaker sessions will be further incorporated in the future and will be linked to assessment learning outcomes.

The right combination of professionalism, friendliness, humour and motivation triggers the majority of students' interest and creates the right atmosphere where students feel comfortable with sharing their ideas.

A surprising – in a very positive way – experience was that students were expecting the tutor in class (online and face) 10 minutes before the starting time for every seminar and were always prepared for the seminar activities.

The class debate was entirely student led. On one occasion, the students had a debate practice outside the classroom on their own initiative, with the view to enhancing their experience and maximizing the outcome of the debate.

The "classroom culture" approach aims to lead to high-satisfaction rates and the creation and maintenance of very good and authentic relationships with students after graduation. Something that can be invaluable for multiple reasons (student ambassadors in their home countries, alumni events, guest speaker presentations, international field trips, international internships, etc.). The authors' personal experience has been very positive.

The classroom culture approach has been tested on online learning environments along with face to face. Evidence suggests that it can be applied well on both educational settings. Students do have very similar learning needs and the element of engagement and involvement can be seen as an essential one. More data and evidence will allow the researcher to generalise findings and come up with practical suggestions and observed outcome. This will inform further own research on the development of an ideal teaching and learning approach in online education.

6.6   POTENTIAL LIMITATIONS

A quantitative research method involves structured questionnaire with close-ended questions. It leads to limited outcomes outlined in the research proposal. So, the results cannot always represent the actual occurring in a generalized form. Also, the respondents have limited options of responses, based on the selection made by the researcher. Another potential bias is the prospect of bias in the interpretation of the results. Although positivism encourages researchers to disregard human emotion and behaviour, there is no guarantee that this will occur at all times during the study. To eliminate bias, the researcher will make sure to develop the research instrument from multiple as well as contrasting theoretical positions to help mitigate interpretative biases.

6.7   RELIABILITY AND VALIDITY

Bolarinwa (2015) advises on the importance of the reliability and validity in research studies, especially in those carried out in developing countries. It has been observed that such confirmation is not a common practice when social and health science research studies are carried out in those countries.

Validity refers to the degree which a measurement measures the purpose that the research is designed for (Bolarinwa, 2015) and thus, refers to the accuracy of a measurement.

On the other hand, still according to Bolarinwa, reliability refers to the replicability of the results collected and means how stable the scores will be over time or across respondents.

6.8   ORIGINALITY/VALUE

It is proposed that each of the engagement dimensions can have both a "positive" and a "negative" pole, each of which represents a form of engagement, separated by a gulf of non-engagement (withdrawal or apathy). The terms "positive" and "negative" are used here not to denote value judgment, but rather to reflect the attitude of the learner.

COVID-19 has changed the way that students learn across the world. Online education is the new normal in many cases in higher education. The common objective is to maximize students' learning experience in the contemporary virtual learning environments. The originality aspect of this working paper is to explore and combine the different engagement dimensions, especially cognitive engagement teaching and learning strategies, and suggests practices that will lead to a high student engagement and satisfaction rates in the contemporary, rapidly changing, diverse environment of higher education.

6.9   CONCLUSION

Research has demonstrated that engaging students in the learning process increases their attention and focus, motivates them to practice higher-level critical thinking skills, and promotes meaningful learning experiences. Strategies include, but are not limited to, brief question-and-answer sessions, breaking news, scenario-based learning, discussion integrated into the lecture, self-discovery activities, ongoing reflection, impromptu, hands-on activities, and experiential learning events. As the instructor adopts the classroom culture approach, they should consider ways to set clear expectations, design effective evaluation strategies and provide ongoing, constructive feedback. Most importantly, placing the learner at the heart of the learning process will signify a much-needed shift of power in the teacher-student relationship. The student will have a strong voice, a choice in learning, enabled to learn what is relevant for them in ways that are appropriate. This is more vital than ever before in the post-COVID era to help students form close and caring relationships with their teachers and peers, fulfil their emotional need for a connection with others and a sense of belonging in society. The pandemic has deprived students of their sense of social belonging, made them more passive and detached them from their own learning experiences. The classroom culture approach suggests an open, empathic and student-centred approach where the student is seen as an "equal partner in education" – respected, listened to, empowered, effectively a co-creator in the learning process.

REFERENCES

ACER (2008) Attracting, Engaging and Retaining: New Conversations about Learning. Australasian Student Engagement Report. Camberwell, Victoria: Australian Council for Educational Research.

Ahlfeldt S, Mehta S and Sellnow T (2005) Measurement and analysis of student engagement in university classes where varying levels of PBL methods of instruction are in use, Higher Education Research and Development 24(1): 5–20.

Ainley M (2006) Connecting with learning: Motivation, affect and cognition in interest processes, Educational Psychology Review 18: 391–405.

Appleton, JJ, Christenson, SL, & Furlong, MJ (2008). Student engagement with school: Critical conceptual and methodological issues of the construct. Psychology in the Schools 45(5): 369–386. http://dx.doi.org/10.1002/pits.20303

Bada SO and Olusegun S (2015) Constructivism learning theory: A paradigm for teaching and learning, Journal of Research & Method in Education 5(6): 66–70.

Barnett R and Coate K (2005) Engaging the Curriculum in Higher Education. Maidenhead: Society for Research into Higher Education and Open University Press.

Baron, P, & Corbin, L (2012). Student engagement: Rhetoric and reality. Higher Education Research & Development 31(6): 759–772. http://dx.doi.org/10.1080/07294360.2012.65 5711

Ben-Eliyahu, A, Moore, D, Dorph, R, & Schunn, CD (2018). Investigating the multidimensionality of engagement: Affective, behavioral, and cognitive engagement across science activities and contexts. Contemporary Educational Psychology 53: 87–105. https://doi.org/10.1016/j.cedpsych.2018.01.002

Bolarinwa OA (2015) Principles and methods of validity and reliability testing of questionnaires used in social and health science researches, Nigerian Postgraduate Medical Journal 22(4): 195–201. doi: 10.4103/1117-1936.173959.

Dunne E and Owen D (2013) The Student Engagement Handbook: Practice in Higher Education Hardcover. United Kingdom: Emerald

Enhancing the Quality of the Student Experience 2016, Inside Government.

Enhancing the Quality of the Student Experience Forum 2015. London: Inside Government.

Finn, JD, & Zimmer, KS (2012). Student engagement: What is it? Why does it matter? In *Handbook of research on student engagement* (pp. 97–131). Springer. http://dx.doi.org/10.1007/978-1-4614-2018-7_5

Fredricks, JA, Blumenfeld, PB, Friedel, J, & Paris, A (2004) *A increasing engagement in urban settings: An analysis of the influence of the social and academic context on student engagement.* April paper presented at the annual meeting of the American Educational Research Association New Orleans Google Scholar

Fredricks, JA, Filsecker, M, & Lawson, MA (2016). Student engagement, context, and adjustment: Addressing definitional, measurement, and methodological issues. Learning and Instruction 43: 1–4. https://doi.org/10.1016/j.learninstruc.2016.02.002

Gayle G and Kaufeldt M (2015) The Motivated Brain: Improving Student Attention, Engagement, and Perseverance, Association for Supervision & Curriculum Development.

HEFCE and BIS on implementing the White Paper for teaching excellence 2016, Inside Government.

Hockings C, Cooke S, Yamashita H, McGinty S and Bowl M (2008) Switched off? A study of disengagement among computing students at two universities, Research Papers in Education 23(2): 191–201.

Howe K (2004) A critique of experimentalism, Qualitative Inquiry 10(1): 42–61.

Improving Student Experience and Engagement 2017, Understanding Modern Government.

Johnson D, Soldner M, Leonard J, et al. (2007) Examining sense of belonging among first-year undergraduates from different racial/ethnic groups, Journal of College Student Development 48(5): 525–542.

Kiernan E, Lawrence J and Sankey M (2006) Preliminary essay plans: Assisting students to engage academic literacy in a first year communication course. Paper presented at the 9th Pacific Rim First Year in Higher Education Conference: Engaging Students. Griffith University, Gold Coast, Australia, 12–14 July.

Krause K-L (2005) Engaged, inert or otherwise occupied? Deconstructing the 21st century undergraduate student. Keynote paper presented at the Sharing Scholarship in Learning and Teaching: Engaging Students Symposium, James Cook University, Townsville, 21–22 September.

Krause K-L and Coates H (2008) Students' engagement in first-year university, Assessment and Evaluation in Higher Education 33(5): 493–505.

Kuh, GD, Cruce, TM, Shoup, R, Kinzie, J, & Gonyea, RM (2008). Unmasking the effects of student engagement on first-year college grades and persistence. The Journal of Higher Education 79: 540–563. https://doi.org/10.1353/jhe.0.0019

Ma, J, Han, X, Yang, J, & Cheng, J (2015). Examining the necessary condition for engagement in an online learning environment based on learning analytics approach: The role of the instructor. Internet and Higher Education 24: 26–34. https://doi.org/10.1016/j.iheduc.2014.09.005

Reeve, J, & Tseng, C-M (2011). Agency as a fourth aspect of students' engagement during learning activities. Contemporary Educational Psychology 36(4): 257–267. doi: 10.1016/j.cedpsych.2011.05.002

Schuetz P (2008) A theory-driven model of community college student engagement, Community College Journal of Research and Practice 32: 305–324.

Skinner, EA, & Pitzer, JR (2012). Developmental dynamics of student engagement, coping, and everyday resilience. In SL Christenson, AL Reschly, & C Wylie (Eds.), *Handbook of research on student engagement* (pp. 21–44). Springer Science + Business Media. https://doi.org/10.1007/978-1-4614-2018-7_2

Trowler V (2010) Student engagement literature review. Department of Educational Research Lancaster University.

Umbach PD and Wawrzynski MR (2005) Faculty do matter: The role of college faculty in student learning and engagement, Research in Higher Education 46(2): 153–184.

Wenger E (1988) Communities of Practice-Learning, Meaning and Identity. United Kingdom: Cambridge University Press.

Yorke M (2006) Student engagement: Deep, surface or strategic? Keynote address delivered at the 9th Pacific Rim First Year in Higher Education Conference: Engaging Students. Griffith University, Gold Coast, Australia, 12–14 July.

Yorke M and Knight P (2004) Self-theories: Some implications for teaching and learning in higher education, Studies in Higher Education 29(1): 25–37.

Zepke N and Leach L (2005) Integration and adaptation: approaches to the student retention and achievement puzzle, Active Learning in Higher Education 6(1): 46–59.

Zhao C and Kuh G (2004) Adding value: Learning communities and student engagement, Research in Higher Education 45(2): 115–138.

7 Human Resource Management in Education

A Brief Overview

Daniel Moreira[1] and Carolina Feliciana Machado[1,2]
[1]School of Economics and Management,
University of Minho, Braga, Portugal
[2]Interdisciplinary Centre of Social Sciences (CICS.
NOVA.UMinho), University of Minho, Braga, Portugal

CONTENTS

7.1 INTRODUCTION

According to Byars and Rue (2006), human resource management is defined as a set of policies, practices and strategies focused on the successful management of an organization's employees in order to achieve its goals. That is the importance rests on the organization's efficiency in pursuing its purposes. Its principles are based on the rules of recruitment, staff management, well-being, maintenance, training, placement, promotion, motivation, compensation, transfer and discipline of the workforce (Omebe, 2014). The same logic applies to the education system, where it acquires the organizational role and teachers/educators are considered the driving force that allows the objectives of education to be fulfilled. In the same way, if this force is not motivated or is duly able to meet the standards of the education system, proper student development will not be possible. Thus, this brief review of the existing literature will be based on three distinct parts: (1) An overview of the different human resource management policies that have been introduced in the education system of countless countries around the world; (2) The challenges that tend to persist and that have been identified by several

DOI: 10.1201/9781003021230-7

authors; (3) The recommendations that some authors make for the possible resolution of problems inherent to the management of human resources in education.

7.2 HUMAN RESOURCE MANAGEMENT IN EDUCATION

It is important to consider the heterogeneous nature of education systems between different countries, the way in which they are organized differs and the *modus faciendi* too. So, trying to adapt a global study to a policy that depends on national governments becomes challenging. However, for this brief literature review, it will be considered the similarities that exist between educational systems, since regardless of the country and the specificity of its education, all educational processes depend on teachers to carry out their programmes.

According to Omebe (2014), the management of human resources in education is essentially based on three aspects: (1) determining the need for staff; (2) satisfy the need for staff and (3) manage and improve staff services. When managed in the best way, human resources in the education system represent a source of competitive strength for the national market of the future, thus, their strategic role cannot in any way be ignored by governments and respective public entities. We can therefore consider that to meet the three aspects mentioned earlier, those responsible for the education system should look for professionals who meet the needs of schools, whether they are teachers/educators or employees in other areas (such as canteens), as well as attending to the effective capabilities of professionals who aim to hire. That is, the professional's degree of professionalization, because they will have a spill-over effect on the student. Referring Omebe (2014), the functions of those who manage human resources in the education system will be: (1) to maintain the staff, (2) pay attention to the relationships between the staff, (3) develop the staff, (4) hiring staff and (5) reward professional performance.

With regard to staff maintenance, the focus is on making the work environment conducive to workers. In other words, the concern is to guarantee fair and clear norms of promotion/transfer, motivation, safety and health (Hdiggui, 2006). As for staff relations, a good communication network at school is essential, something that allows workers to be aware of the progress being made and that makes them feel included in the decisions taken (Omebe, 2014).

In staff development, the key is in the process of identifying flaws, within the scope of performance evaluation, which it contains. This is something that can sensitize management to provide the necessary training to improve skills, as the success or failure of school goals is also dependent on the abilities of those who teach. The beginning of the human resource management functions starts with hiring staff (Omebe, 2014). As far as schools are concerned, this work is performed (usually) by the Ministry of Education. The latter is responsible for obtaining people with the skills and knowledge necessary for the proper performance of the functions, while the administration must, in turn, reward those who really stand out in their performance. As a rule, the Ministries of Education do not pay due attention to this last point. In fact, they tend to base their policies on administrative and financial principles (Hdiggui, 2006), not decentralizing decision-making power.

7.3 CHALLENGES

Human resource management is in itself the biggest challenge for an organization (whether public or private), something that is proving to be a trend that will not change in the coming years (Bucata, 2018). The demographic shifts that have taken place in both the developed and the developing worlds are putting pressure on both government and the private sector to initiate and implement creative solutions that educate, integrate and sustain a rapidly changing and diverse working population. Human resource management has become remarkably complex in the sense that, as human beings, they are not expected to do something over and over again in the same way (Omebe, 2014). We can infer that their productivity is then dependent on numerous variables that, at times, even prove to be uncontrollable. Thus, there are numerous challenges that appear in the process of managing human resources in education. In this chapter, it will be highlighted three factors that generate setbacks in the 21st century, namely (Omebe, 2014): (1) poor working conditions; (2) the problems relating to staff, and the staff problems and (3) the constant assumption of the use of information technologies.

First, the poor working conditions slip, from the outset, into the remuneration policy. If a good remuneration policy is in place, it will reduce inequalities among workers, increase their morale, and will motivate them to do their best in their roles with a view to promotion or a reward. Now, it is known that in some countries, mainly in developing countries, teachers receive their salaries months late (e.g. South Africa), and tend not to claim their rights for fear of reprisals (Tshehle, 2016).

However, the poor conditions are not only revealed in salaries. Also, the little investment made in schools tends to affect the work of those who teach, as well as problems with recruitment that often does not meet the required quantity or quality (Omebe, 2014). Another problem is the growth of indiscipline in schools, both among students and staff, with consequences that are further aggravated by little or no supervision of existing human resources.

Dissatisfaction is growing, and the phenomenon of teacher transfers is recurrent, since for convenience teachers prefer to remain in urban areas, while the needs of rural areas are not met, and if they are, it is through government orders. Adopting a critical stance, the increase in teachers' salaries does not solve all the adversities that were mentioned. In fact, if we take Portugal as an example, teachers are among the highest paid civil servants when compared to those who had higher education. However, they face the misfortune of having an overload of work, of facing a huge amount of bureaucracy and not being, in any way, recognized for their work, reason why they are among the most dissatisfied teaching staff in the entire European Union (Leiria, 2016). Still referring that, in the 21st century, constant contact with information technologies is required, it is useful, in fact. However, the investment in training the skills of teachers with these technologies is, for the most part, non-existent. Now, the technology that should be a facilitator of teaching becomes a conditioning factor.

In short, it is above all the lack of follow-up, and in a way the idea that their opinion is not considered when making decisions, that contributes to the stagnation of the teachers' motivation. Added to this, we can refer to the low mobility of careers and the lack of support from the directions and administrations (be it because of cuts in funds or because of administrative requirements) at the time of teaching (Hdiggui, 2006).

7.4 RECOMMENDATIONS

The authors (Hdiggui, 2006; Omebe, 2014) consider the existence of human resource management in education to be essential and, therefore, make some recommendations on how the work environment of teachers and students can be improved. First, education has to be more attractive to those who teach, something that can only happen by creating a conducive atmosphere in schools. Then, governments must pay more attention to the implementation of policies in the education sector, as this is the basis for progress in other sectors of society. A fair salary structure must be implemented in order to comply with the functions performed by the teacher, never failing to meet the same category, thereby differentiating the assigned values. It is necessary to pay attention to "computer literacy", this responding to the phenomenon of globalization in which we live in the 21st century and enhancing educational productivity. In developing countries, education must be standardized and must contain a high level of teaching on social norms so that it contributes to the socio-economic development of the countries concerned (Omebe, 2014).

High-quality teaching is more important than anything else schools do. Thus, it would be possible to assume that states, and schools, would focus on attracting, training and supporting the best people to achieve educational goals. In other words, it is important to attend to the quality of those who teach and, however, this is traditionally measured through the possession of degrees or other academic degrees, something that in itself does not offer us clues about who will be more efficient in student development (Olson, 2008). Without data systems that can track individual teachers over time, tracking their performance based on the preparation programme they graduated from or how successful their students are is virtually impossible. Building a stronger data infrastructure is one of the most basic steps ministries of education can take to design a better human capital system in education (Olson, 2008).

To combat the phenomenon of resistance to the transfer of teachers to rural or less desirable areas, administrations must put in place a set of financial incentive policies for the establishment of teachers in these areas, as well as investing in the improvement of schools in the target regions. Additionally, incentives such as investing in improving the range of knowledge/capacities of teachers while they are exercising their functions can contribute to fostering teachers' motivation and confidence in the system. Studies show that teachers do not come to classrooms ready to teach, contrary to popular belief, they are practically subjected to knowledge based on the experience they obtain in the first years of teaching.

Another concern will be to meet the psychological needs of teachers, who are not properly monitored in their work. In certain situations, teachers placed in schools with a critical/at-risk environment, are practically left to be abandoned, not having a close entity to turn to in an extreme situation. A teacher support office may be one of the measures (Hdiggui, 2006).

The implementation of these recommendations presupposes a greater decentralization of decision-making, so that each school administration can take the most appropriate action to the environment that it has in its school.

7.5 CONCLUSION

In short, introducing human resource management in the education sector will benefit the society in question, and renew the motivation of all those who are an integral part of the system. The key to its realization is to gain quantitative and qualitative knowledge of the existing set of human resources, their needs in the short and medium term, as well as the supply and demand for skills in the labour market (Hdiggui, 2006). Education is no different from a company, and the challenges that arise are as or more challenging than those that appear in a corporate logic. According to Barros (2014), in order to answer to all these challenges, human resources departments must adopt strategies compatible with the professional development of employees, through the correct management of flexible work. This is critical considering that we are facing a new workforce, which is more autonomous, works in a network system, but which at any time can change its professional trajectory depending on its motivations and/or the organization's needs. This serves to demonstrate the primacy that education has for the economic engine of a country, and to alert the respective ministries of education to the attention they must pay to the management of their human resources. The world is changing, the phenomenon of globalization is current, hence the need to introduce methodological reforms in an education system which, let us be honest, has remained practically unchanged over time. It is necessary to raise awareness among administrative decision-makers, so that both those who teach and those who are taught contribute to achieving the goals that the "school" proposes.

REFERENCES

Barros, Cristina (2014). Recursos Humanos no século XXI: desafios e oportunidades. RH Magazine, https://rhmagazine.pt/recursos-humanos-no-seculo-xxi-desafios-e-oportunidades/, accessed in 18 December 2021.

Bucata, G. (2018). The challenges of human resources department – the impact of the demographic evolution (the case of migration). Sciendo, Scientific Bulletin, Vol. XXIII, No. 2(46), pp. 81–88.

Byars, Lloyd and Leslie W. Rue (2006). Human Resource Management. McGraw-Hill/Irwin, Ed. 8.

Hdiggui, El Mostafa (2006). Human Resource Management in the Education Sector. Unesco, http://unesdoc.unesco.org/images/0015/001508/150801e.pdf, accessed in 15 December 2021.

Leiria, Isabel (2016). Professor português: salário acima da média, muito trabalho, pouco reconhecimento. Expresso, https://expresso.pt/sociedade/2016-10-17-Professor-portugues-salario-acima-da-media-muito-trabalho-pouco-reconhecimento, accessed in 15 December 2021.

Olson, Lynn (2008). Human Resources a Weak Spot. Education Week. http://www.edweek.org/ew/articles/2008/01/10/18overview.h27.html, accessed in 18 December 2021.

Omebe, Chinyere (2014). Human resource management in education: issues and challenges. British Journal of Education, Vol. 2, No. 7, pp. 26–31.

Tshehle, Boitumelo (2016). Teachers not paid salaries. http://www.sowetanlive.co.za/news/2016/04/28/teachers-not-paid-salaries, accessed in 18 December 2021.

Index

For Product Safety Concerns and Information please contact our EU
representative GPSR@taylorandfrancis.com
Taylor & Francis Verlag GmbH, Kaufingerstraße 24, 80331 München, Germany